Ann-Karin Sanchez

W and Z Bosons at the LHC

Ann-Karin Sanchez

W and Z Bosons at the LHC

Detailed studies with the CMS detector

Südwestdeutscher Verlag für Hochschulschriften

Impressum / Imprint
Bibliografische Information der Deutschen Nationalbibliothek: Die Deutsche Nationalbibliothek verzeichnet diese Publikation in der Deutschen Nationalbibliografie; detaillierte bibliografische Daten sind im Internet über http://dnb.d-nb.de abrufbar.
Alle in diesem Buch genannten Marken und Produktnamen unterliegen warenzeichen-, marken- oder patentrechtlichem Schutz bzw. sind Warenzeichen oder eingetragene Warenzeichen der jeweiligen Inhaber. Die Wiedergabe von Marken, Produktnamen, Gebrauchsnamen, Handelsnamen, Warenbezeichnungen u.s.w. in diesem Werk berechtigt auch ohne besondere Kennzeichnung nicht zu der Annahme, dass solche Namen im Sinne der Warenzeichen- und Markenschutzgesetzgebung als frei zu betrachten wären und daher von jedermann benutzt werden dürften.

Bibliographic information published by the Deutsche Nationalbibliothek: The Deutsche Nationalbibliothek lists this publication in the Deutsche Nationalbibliografie; detailed bibliographic data are available in the Internet at http://dnb.d-nb.de.
Any brand names and product names mentioned in this book are subject to trademark, brand or patent protection and are trademarks or registered trademarks of their respective holders. The use of brand names, product names, common names, trade names, product descriptions etc. even without a particular marking in this work is in no way to be construed to mean that such names may be regarded as unrestricted in respect of trademark and brand protection legislation and could thus be used by anyone.

Coverbild / Cover image: www.ingimage.com

Verlag / Publisher:
Südwestdeutscher Verlag für Hochschulschriften
ist ein Imprint der / is a trademark of
OmniScriptum GmbH & Co. KG
Heinrich-Böcking-Str. 6-8, 66121 Saarbrücken, Deutschland / Germany
Email: info@svh-verlag.de

Herstellung: siehe letzte Seite /
Printed at: see last page
ISBN: 978-3-8381-3622-6

Zugl. / Approved by: Zurich, ETH Zurich, Diss., 2012

Copyright © 2013 OmniScriptum GmbH & Co. KG
Alle Rechte vorbehalten. / All rights reserved. Saarbrücken 2013

Abstract

W and Z boson production has been studied in proton-proton collisions at the Large Hadron Collider at a center of mass energy of $\sqrt{s} = 7\,\text{TeV}$. A nearly background free event selection is achieved after identifying the bosons from their decays to high energy leptons measured with the Compact Muon Solenoid detector and applying additional selection cuts.

The central charged-particle multiplicity and the forward energy flow as well as correlations between them are analyzed in detail with a data sample corresponding to an integrated luminosity of 36 pb^{-1}. These observables are sensitive to the underlying event structure. However, the underlying event is still poorly understood and thus has to be described with parametrized models. Experimental data is used to tune and constrain the models. Measurements with W and Z bosons are especially useful for the further tuning of the existing models, due to a clear separation of the hard interaction from the multiparton interactions. The measured distributions are compared to the different tunes of the available simulations. It is found that none of them describes the observations and the data can thus be used to constrain and improve the models.

A subsample of events with a pseudorapidity gap of at least 1.9 units in the forward direction is studied. Such a large rapidity gap is a common signature for diffractive events, and it is found that $(1.46 \pm 0.09\ (\text{stat.}) \pm 0.38\ (\text{syst.}))\%$ in W events and $(1.57 \pm 0.25\ (\text{stat.}) \pm 0.42\ (\text{syst.}))\%$ in Z events have a gap signature. It is shown that, as a result of inaccurate descriptions by models, the rapidity gap signature is not sufficient to provide evidence for a diffractive component in the data. However, diffractive events could be confirmed with an asymmetry variable, showing that the majority of the leptons are found in the hemisphere opposite to the gap. By fitting the shape of a simulated sample with a variable fraction of diffractive and non-diffractive events to data, a diffractive component of $(50.0 \pm 9.3\ (\text{stat.}) \pm 5.2\ (\text{syst.}))\%$ is determined.

Finally, a precise measurement of the W and Z boson transverse momenta is presented.

Through the reconstruction of the Z boson via the lepton four-vector, accurate results are obtained for the Z momentum. Z+jet events are used to control and validate the W transverse momentum through the recoiling jet in W+jet events. The results are presented for a transverse momentum up to 500 GeV for inclusive events and events with one or two jets for an integrated luminosity of 1.5 fb^{-1}. The distributions can potentially be used to constrain the gluon distribution function to high precision at LHC energies.

Contents

Abstract		i
1 Introduction		**1**
2 W and Z Boson Physics at the LHC		**3**
2.1	The Standard Model of Particle Physics	3
	2.1.1 Electroweak Theory	5
	2.1.2 Quantum Chromodynamics	9
2.2	Proton-Proton Collisions	11
	2.2.1 Factorization and Parton Distribution Functions	11
	2.2.2 W and Z Boson Production and Decay Channels	13
	2.2.3 Jets in W and Z Events	16
	2.2.4 Underlying Event	18
	2.2.5 Diffractive Processes	19
2.3	Monte Carlo Event Generators	20
3 The CMS Experiment at the LHC		**25**
3.1	The Large Hadron Collider	25
3.2	The CMS Detector	29
	3.2.1 Overview	29
	3.2.2 Magnet	30
	3.2.3 Inner Tracking System	31
	3.2.4 Electromagnetic Calorimeter	32
	3.2.5 Hadron Calorimeter	35
	3.2.6 Muon System	36
	3.2.7 Trigger	38
	3.2.8 Computing	40
3.3	CMS Physics Measurements	41

    3.3.1   Data Taking Periods 2010 and 2011 of CMS at LHC 41
    3.3.2   CMS Measurements with W and Z Bosons 42

4 Muon Identification   47
  4.1 Muon Reconstruction Algorithms 48
  4.2 Muon Momentum Resolution 49
  4.3 Muon Selection 52
    4.3.1   Muon Isolation and Quality Selection Cuts 54
    4.3.2   Muon HLT Triggers 58

5 W and Z Boson Event Selection   59
  5.1 Missing Transverse Energy 59
  5.2 Jets 62
  5.3 Vertex Reconstruction 63
  5.4 W and Z Signal Selection 64
    5.4.1 $Z \to \mu^+\mu^-$ Signal Selection 64
    5.4.2   Results with $\sqrt{s} = 7\,\text{TeV}$ Data 66
    5.4.3 $W \to \mu\nu$ Signal Selection 73
    5.4.4   Results with $\sqrt{s} = 7\,\text{TeV}$ Data 74
    5.4.5 $Z \to e^+e^-$ and $W \to e\nu$ Signal Selection 77

6 Diffractive processes in W and Z events   83
  6.1 Event Topology 84
  6.2 Pileup Influence and Rejection 85

7 Forward Energy Flow and Central Charged-Particle Multiplicity   91
  7.1 Results for the Forward Energy and Central Multiplicity Measurements . . 91
  7.2 Soft Pileup Events and Forward Energy Distributions 96
  7.3 Correlations 97
  7.4 Interpretation of the Observed Correlations 100
  7.5 Corrections to Hadron Level 103
    7.5.1   Cuts on Generator Level 103
    7.5.2   Bin-by-bin Correction Factors 105
    7.5.3   Response Matrix 109

8 W and Z Events with a Large Pseudorapidity Gap   113
  8.1 Observed Number of LRG Events 113
  8.2 Jet Activity and Exclusive W/Z Production 115

8.3	Size of the Pseudorapidity Gap and Central Gaps	116
8.4	Charged-Particle Multiplicity and Forward Energy	117
8.5	Hemisphere Correlations Between the Gap and the W (Z) Boson	119

9 W and Z Transverse Momentum Spectrum 123
    9.1 Transverse Momentum and PDFs 124
    9.2 Transverse Momentum Measurement 127
        9.2.1 Results for Z p_T 129
        9.2.2 Results for W p_T 134

10 Conclusions 139

A Diffractive Processes: Additional Event Displays 141

B Forward Energy and Particle Multiplicity: $Z \to \mu^+\mu^-$ distributions 145

C W and Z p_T Ratio Measurements 149

List of Tables 151

List of Figures 153

Bibliography 157

Acknowledgments 167

Chapter 1

Introduction

On 23 November 2009 the first proton collisions from the LHC were recorded with the CMS detector. Many years have passed since the early planing phase of the machine, and now there is great hope in gaining new perspectives on our insights of the fundamental forces and particles. Even though a remarkable understanding has been developed, many questions remain unanswered. The LHC and the detectors were designed to provide answers on many aspects and the technology has been pushed to the extreme to achieve this. One of the most prominent is the quest for the missing piece providing the masses of particles. A promising candidate is the Higgs Boson and the experiments will soon provide definitive evidence if it exists or not. Still, many other open questions remain. There is no evidence so far, of what the nature of dark matter is, why there is an asymmetry between matter and antimatter in the universe, or how gravity can fit in the whole concept.

The experiments at the LHC are addressing many of these questions. Before exploring new realms however, one of the first things which can be accomplished is to confirm the established physics at the new LHC energy scale. Extrapolations of the physics measured at lower energies deliver predictions for the higher energy regime. The verification of these predictions settles the *Standard Model of Particle Physics*, and serves as input for other predictions. Many parameters of the Standard Model have to be determined by experimental measurements and some phenomena even rely entirely on phenomenological models. Thus, the precise measurement of distributions sensitive to them serve to either validate and improve the current models, or to detect inconsistencies which indicate the presence of new physics.

In this thesis, these points are addressed in the context of measurements with W or Z bosons produced in the collision and identified by their decay to muons. First, a phenomenon called *diffractive process* is studied. These kind of processes make up a considerable fraction of the events at the LHC but are not yet well understood. The detailed study of such events can be used to gain a deeper understanding of the structure and the dynamics. The analysis of diffractive events is closely related and influenced by other underlying physics, due to the composite nature of protons. Two variables which are sensitive to the underlying physics are the total energy flux and the multiplicity of particles detected in the experiment. They are extensively studied and compared to existing models. In a second study, the momentum transverse to the proton beam axis of W and Z bosons is analyzed. The precise measurement can be used to constrain certain quantities, such as the so-called *Parton Distribution Functions*, which are not predicted by theory and rely on experimental determination.

First, the theoretical and phenomenological framework relevant to this work are introduced in chapter 2. The detector used for the analysis is described in detail in chapter 3. In chapter 4 the selection and identification of muons is elaborated, followed by the W and Z event selection in chapter 5. There, the results with the CMS data are shown, providing the basis for the subsequent studies. In chapters 6 to 8 the diffractive analysis as well as the study of the forward energy flow and the central charged-particle multiplicity are presented. Finally, the transverse momentum analysis is described in chapter 9.

The diffractive studies presented in this thesis were published [1] as a combined result with the vector-bosons decaying to electrons. Some of the combined measurements will be shown here, to illustrate the fact that equal results are obtained and especially when the result becomes more significant by doubling the data. A large part of the descriptions is taken from the publication [1].

Throughout this thesis some conventions will be used for ease of reading. Most of the time when speaking of a particle, the anti-particle is meant as well. The indices specifying the charge and the flavor of the neutrino are usually omitted. For simplicity, natural units are used, with $c = \hbar = 1$, and the energy and momentum are given in electronvolts (eV). A 'transverse' variable is always transverse to the axis of the proton beam. The rapidity $y = \text{artanh}(v/c)$ is approximated by $y = \frac{1}{2} \ln \frac{E+p}{E-p}$. In the massless limit this can be interchanged with the pseudorapidity $\eta = \frac{1}{2} \ln \frac{|\vec{p}|+p}{|\vec{p}|-p}$, which is used as a lorentz invariant alternative to the polar angle θ to specify the angle relative to the beam axis. For muons and electrons mostly the pseudorapidity is used, while for the heavier W and Z bosons this approximation is not valid.

Chapter 2

W and Z Boson Physics at the LHC

In this chapter, the theoretical and phenomenological framework relevant to this work are introduced. The foundations for the standard model of particle physics are laid with the quantum field theories, describing the fundamental interactions and building blocks of particle physics. Phenomenological models complement where the theoretical descriptions fail. The link from the theoretical and phenomenological models to experimental data can be done with simulations. These provide a tool to translate the theoretical predictions to expected event properties, and from measured data to model verification.

2.1 The Standard Model of Particle Physics

The Standard Model of Particle Physics (SM) is a theory which provides an excellent description of the observed particles and interactions. It is a relativistic quantum field theory, based on a local $SU(3)_C \times SU(2)_L \times U(1)_Y$[1] gauge symmetry. It describes three of the four known fundamental forces: the electromagnetic, the weak, and the strong interaction between fundamental matter. The electromagnetic and the weak force are described by one unified theory, the electroweak theory and the strong interaction between quarks and gluons by Quantum Chromodynamics (QCD). The SM particles are classified in 12 matter particles (fermions) and the particles associated to the interaction field (gauge bosons). The matter particles consist of 6 leptons and 6 quarks and are arranged in three almost identical families except for their masses, which increase from the first to the third family. Due to this, the particles of the first family with the lowest mass cannot

---
[1] C refers to the color charge related to the strong interaction, L to the weak isospin coupling to left-handed fermions and Y to the hypercharge related to the electroweak interaction

decay and form the building blocks of observed stable matter. The 6 leptons are the electron, the muon and the tau with their corresponding neutrinos. The neutrinos only carry the charge coupling to the electroweak force while the other 3 leptons carry electric charge and thus interact electromagnetically. The quarks carry an additional charge, the color which couples to the strong force, and are the only matter particles coupling to all forces. The color charge can explain why quarks are only observed in bound states forming hadrons. These can be mesons, formed by quark-antiquark pairs, or baryons, formed by three quarks or antiquarks. To each fundamental force described in the SM a corresponding gauge boson has been found. The photon is the exchange particle of the electromagnetic force, the Z and $W^{\pm}$ bosons of the weak force and the gluons are associated to the strong force. While the photon and the gluons are massless, the W and Z bosons are very heavy. Due to this the range of the electromagnetic force is infinite whereas the weak interaction is short-ranged. The massless gluons however, are subject to their own force since they carry color charge which results in the confinement of quarks and gluons and thus a restricted range of the strong force. Table 2.1 shows an overview of the SM building blocks with the corresponding masses, electric charges and spins.

Table 2.1: Overview of the building blocks of the Standard Model of Particle Physics. A complete review can be found in [2].

		Standard Model of Particle Physics					
		2.4 MeV u 2/3 1/2 up	1.27 GeV c 2/3 1/2 charm	171.2 GeV t 2/3 1/2 top		0 γ 0 1 photon	mass charge spin name
Fermions	Quarks	4.8 MeV d -1/3 1/2 down	104 MeV s -1/3 1/2 strange	4.2 GeV b -1/3 1/2 bottom	Bosons	0 g 0 1 gluon	
	Leptons	< 2.2 eV ν_e 0 1/2 electron neutrino	< 0.17 MeV ν_μ 0 1/2 muon neutrino	< 15.5 MeV ν_τ 0 1/2 tau neutrino		91.2 GeV Z^0 0 1 Z boson	
		0.511 eV e -1 1/2 electron	105.7 MeV μ -1 1/2 muon	1.777 GeV τ -1 1/2 tau		80.4 GeV $W^{\pm}$ ±1 1 W boson	

2.1.1 Electroweak Theory

The electroweak theory describes the electromagnetic and the weak interaction as one single force, merely manifesting differently at low energies. Above the unification energy of around 100 GeV they merge into one single force, the electroweak interaction. It is a spontaneously broken gauge theory formulated by Glashow, Weinberg and Salam (GWS) [3, 4, 5] which up to now has been able to describe and predict the experimentally measured phenomena concerning the electroweak theory. The neutral currents were observed first in neutrino scattering by the Gargamelle Collaboration in 1973 [6]. Later, in 1983 the $W^\pm$ and Z bosons were discovered in $p\bar{p}$ collisions at the Super Proton Synchrotron by the UA1 and UA2 collaborations [7, 8, 9, 10], establishing the electroweak theory.

The Lagrangian of the theory is formulated under assumption of transformation invariance under $SU(2)_L \times U(1)_Y$ symmetries, where the hypercharge Y is the generator of the $U(1)_Y$ symmetry while the $SU(2)_L$ is generated by the weak isospin operators. The subscript L denotes the fact that the gauge bosons associated to the $SU(2)_L$ symmetry only couple to the left-handed fermions, which are described by SU(2) doublets, while the right-handed fermions are described by SU(2) singlets.

The electroweak Lagrangian evolves out of quantum electrodynamics (QED), which is based on the request of invariance under local phase rotation (local U(1) symmetry)

$$\psi(x) \rightarrow e^{ie\alpha(x)}\psi(x) \tag{2.1}$$

where ψ is the field identified as the fermion state with definite energy and momentum and e is the coupling constant, the electric charge of the fermion. The QED Lagrangian is given by

$$\mathcal{L}_{QED} = \bar{\psi}(i\gamma^\mu D_\mu - m)\psi - \frac{1}{4}F_{\mu\nu}F^{\mu\nu} \tag{2.2}$$

The first term describes the free fermion field and emerges naturally by requiring one mass parameter for a free relativistic particle satisfying $E^2 = p^2 + m^2$, where $\mathcal{L}$ is invariant under Lorentz transformation. The latter requirement ensures same physical laws for any relativistic observer. A derivative term is needed to generate the momentum and get the dynamics into the equation. The Dirac Lagrangian $\mathcal{L}_{Dirac} = \bar{\psi}(i\gamma^\mu \partial_\mu - m)\psi$ satisfies global but not local invariance under U(1) transformation. To achieve local U(1) symmetry, the gauge covariant derivative $D_\mu \equiv \partial_\mu - ieA_\mu$ is used instead of the normal derivative ∂_μ. A_μ is a vector field, the photon field and is a direct result of the requirement that the covariant derivative D_μ should transform in the same way as ∂_μ under Lorentz transformation. Thus, by imposing local U(1) symmetry to the Lagrangian of fermions, it automatically describes a photon field.

The second term in eq. (2.2) is the kinetic term associated to the field A, which is needed as it is a physical field. With the electromagnetic field tensor given by $F_{\mu\nu} = \partial_\mu A_\nu - \partial_\nu A_\mu$, it also fulfills gauge invariance and satisfies the Maxwell equations.

To obtain a consistent theory at higher orders in perturbation theory the gauge needs to be fixed. Writing out the covariant derivative and adding the gauge fixing term $\mathcal{L}_{gauge} = -\frac{\xi}{2}(\partial_\mu A^\mu)^2$, we obtain

$$\mathcal{L}_{QED} = \mathcal{L}_e + \mathcal{L}_\gamma + \mathcal{L}_{int} \tag{2.3}$$

with

$$\mathcal{L}_e = \bar{\psi}(i\gamma^\mu \partial_\mu - m)\psi \tag{2.4}$$

$$\mathcal{L}_\gamma = -\frac{1}{4}F_{\mu\nu}F^{\mu\nu} - \frac{\xi}{2}(\partial_\mu A^\mu)^2 \tag{2.5}$$

$$\mathcal{L}_{int} = e\bar{\psi}\gamma^\mu A_\mu \psi \tag{2.6}$$

This Lagrangian describes a free fermion field ψ with mass m, a free photon field A_μ and the interaction between the two. We note that the gauge field is a massless vector field with an infinite range and with no self-interaction. Any additional mass term for A would break the imposed symmetry.

The generalization of QED leads to the non-abelian gauge theories and the description of the weak interaction. This is done by extending the request of invariance under local phase rotation to the SU(2) group. As in QED, the derivative term is generalized by the covariant form to bring the global to a local symmetry:

$$D_\mu = \partial_\mu - igA_\mu^a t^a \tag{2.7}$$

Here g is is the gauge coupling, analogous to the electron charge e in QED, and t^a are the generators of the symmetry group. The vector field is replaced by a set of vector fields A_μ^a, one for each generator. The kinetic term for the vector fields is obtained similarly, with the field tensor $F^a_{\mu\nu}$ defined as

$$F^a_{\mu\nu} = \partial_\mu A^a_\nu - \partial_\nu A^a_\mu + g f^{abc} A^b_\mu A^c_\nu \qquad (2.8)$$

where f^{abc} is a set of numbers called *structure constants* and is defined by the commutation relation of the generators

$$[t^a, t^b] = i f^{abc} t^c \qquad (2.9)$$

The Lagrangian looks very similar to the QED Lagrangian

$$\mathcal{L} = \bar{\psi}(i\gamma^\mu D_\mu - m)\psi - \frac{1}{4}(F^a_{\mu\nu})^2 \qquad (2.10)$$

but with the important difference that now it contains self-interaction terms for the vector field which arise from the non-commutativity of the group, i.e. the additional term in eq. (2.8).

The Lagrangians detailed so far can describe fermions and their interaction with massless gauge bosons with or without self-interaction. However, the experimentally measured W and Z bosons are very heavy. Adding a mass term to the Lagrangian would violate the gauge symmetry. The solution is the integration of a concept called *spontaneous symmetry breaking* to the local gauge invariance theory. This spontaneously broken local symmetry allows the gauge bosons to acquire mass while still being constrained by the underlying symmetry. In the SM this integration is achieved by the so-called Higgs Mechanism, which implies the existence of a massive scalar boson. The Higgs Boson has not been discovered yet and is being searched for at present days to confirm the mechanism of mass generation.

The symmetry breaking is obtained by introducing an external scalar field ϕ, a SU(2) doublet with complex components, with the Lagrangian given by

$$\mathcal{L}_\phi = |D_\mu \phi|^2 - V(\phi) - \frac{1}{4} F_{\mu\nu} F^{\mu\nu} \qquad (2.11)$$

where V is the scalar potential of the Higgs field given by

$$V(\phi) = \mu^2 \phi^\dagger \phi + \lambda (\phi^\dagger \phi)^2. \qquad (2.12)$$

with μ a mass parameter and λ a dimensionless positive self-coupling constant. The global minimum of this potential for $\mu^2 > 0$ is at $\phi_0 = 0$. The symmetry can be broken by choosing $\mu^2 < 0$ resulting in non-vanishing, degenerate minima at

$$|\phi_0|^2 = -\frac{\mu^2}{2\lambda} \equiv v^2 \qquad (2.13)$$

where v is the vacuum expectation value (VEV). By expanding the field around this minimum and replacing the new expression for the field in the Lagrangian 2.11, a mass term for the gauge boson appears from the first term in the Lagrangian fulfilling the condition $m^2 = 2g^2\phi_0^2$.

Combining all these elements, the local U(1) symmetry providing the massless photon field coupling to fermions, and the SU(2) gauge symmetry spontaneously broken by the introduction of a scalar field providing massive weak bosons, we finally find the GWS theory of weak interactions. The gauge field tensor of the $SU(2)_L$ group is denoted by $A^a_{\mu\nu}$ with a running over the 3 generators of the group, the gauge field by A^a_μ, and the coupling by g. The gauge field tensor of the $U(1)_Y$ group is denoted by $B_{\mu\nu}$, the gauge field by B_μ, and the coupling by g'. The mass eigenstate fields, which represent the measurable gauge bosons, can then be written as

$$W^\pm_\mu = \frac{1}{\sqrt{2}}(A^1_\mu \mp iA^2_\mu), \qquad (2.14)$$

$$Z^0_\mu = \frac{1}{\sqrt{g^2 + g'^2}}(gA^3_\mu - g'B_\mu) \qquad (2.15)$$

and the massless field orthogonal to Z^0_μ:

$$A_\mu = \frac{1}{\sqrt{g^2 + g'^2}}(g'A^3_\mu + gB_\mu) \qquad (2.16)$$

with the masses

$$m_W = g\frac{v}{2}, \quad m_Z = \sqrt{g^2 + g'^2}\,\frac{v}{2} \quad \text{and} \quad m_A = 0. \qquad (2.17)$$

The massless gauge boson A_μ then couples to the conserved charge

$$Q = T^3 + \frac{Y}{2} \qquad (2.18)$$

which generates the demanded symmetry. The weak hypercharge Y is the corresponding conserved charge of the $U(1)_Y$ symmetry, and the third component of the weak isospin T^3 the charges generating the $SU(2)_L$ symmetry. This construction of the $SU(2)_L \times U(1)_Y$ symmetry accounts for the observed coupling of the W bosons to only left-handed fermions. The change of basis to the mass eigenstates can be expressed by the weak mixing angle θ_w defined by:

$$\begin{pmatrix} Z^0 \\ A \end{pmatrix} = \begin{pmatrix} \cos\theta_w & -\sin\theta_w \\ \sin\theta_w & \cos\theta_w \end{pmatrix} \begin{pmatrix} A^3 \\ B \end{pmatrix} \qquad (2.19)$$

With the relation $g = \frac{e}{\sin\theta_w}$, the terms in the Lagrangian which deliver the couplings of the weak bosons can be rewritten such that all the weak boson couplings depend only on the two parameters, e and θ_w. Additionally, eq. (2.17) gives $m_W = m_Z \cos\theta_w$ and thus, the (tree level) electroweak process can be described by m_W, e and θ_w. Higher order corrections however, depend also on the top mass and the Higgs mass.

2.1.2 Quantum Chromodynamics

The strong interaction is based on the non-abelian SU(3) gauge symmetry. The corresponding conserved charge is the color charge with values conventionally assigned to red, green and blue. The main difference to the electromagnetic force is that the gauge bosons, the gluons, carry charge themselves giving rise to the phenomenon called confinement, the increase of the force between two quarks with increasing distance between them. It can be explained by a gluon string forming due to the color charge, with its energy increasing with growing distance. Hence, at low momentum transfer, the quarks will form hadrons.

The QCD Lagrangian is basically constructed identically as in the previous section. The relevant difference is in the kinetic term of the gauge field and arises from the structure constants f^{abc} defined by the commutation relations of the group. It contains the self-interaction terms of the gluons, triple-gluon and four-gluon vertices and thus, due to the non-commutativity of the group the gauge bosons obtain color.

Quantum field theories rely on perturbation theory as they are usually too complex to calculate an exact solution. QCD is perturbative in high-energy or equivalently short-distance interactions. The so-called ultraviolet divergences which result from infinite momenta for virtual particles in loop radiative corrections can be renormalized by redefining the quark mass and the strong coupling constant α_s such that the infinities are canceled. From the canceling of the ultraviolet divergences a differential equation (renormalization group equation) is obtained for the renormalized (physical) coupling constant.

The solution, which is now finite depends on the energy scale Q:

$$\alpha_s(Q) = \frac{2\pi}{b_0} \frac{1}{\log(Q/\Lambda)} \qquad (2.20)$$

with $b_0 = (11 - \frac{2}{3}n_f)$ and n_f the number of quark flavors with mass smaller than Q and Λ the energy scale above which perturbation theory is valid. This means that for decreasing energies α_s would grow infinitely, which does not happen physically, and thus the theory is not perturbative below a certain scale. The scale Λ is determined from experiments and is approximately 200 MeV. When Q becomes large, α_s diminishes asymptotically and the quarks and gluons composing a hadron can be considered interacting very weakly, i.e. asymptotically free partons.

At the LHC both of these cases will take place in the proton-proton collisions. For the events which are triggered and readout by the detectors, a high momentum transfer between the interacting particles takes place. This results either in the production of a large mass object or in the scattering at a large angle. For these production processes, which happen at a timescale of the order of $\frac{1}{Q}$ perturbation theory can be applied and the cross sections can be calculated. The calculation of the cross sections includes the bare quark or gluon together with the soft and at the same time collinear radiation[2] and its usability for experimental predictions relies on factorization. Due to the confinement phenomenon, at a timescale of $\sim \frac{1}{\Lambda}$ much larger than the production process the quarks and gluons form hadrons, which can not be described in perturbation theory and relies on phenomenological models.

In the following section we will proceed with a more phenomenological approach in describing the physics in proton-proton collisions including the briefly mentioned factorization and hadronization models narrowing down then to the physics of W and Z bosons and phenomena like diffractive processes and others. For more details and calculations on quantum field theories, QCD and particle physics in general these textbooks [11, 12, 13] are good references.

---

[2] Soft is used as synonym to low energetic and collinear refers to the original orientation of the particle.

2.2 Proton-Proton Collisions

2.2.1 Factorization and Parton Distribution Functions

The knowledge of the production cross section of the different SM processes at the LHC is important for any LHC physics program. Yet, the cross section is not only composed of the short-ranged, perturbatively computable partonic cross section but also contains the long-distance phenomena which lead to infrared divergences in the computation. The factorization theorem [14] proves a method to separate the long- from the short-ranged effects by factorizing the process cross section in an incoherent sum of partonic cross sections which are convoluted with functions which absorb the non-perturbative parts. These functions, called parton distribution functions (PDF), provide the probability of finding a parton in the hadron with a certain momentum fraction x and have to be determined experimentally. The functions are the same for any process and can therefore be determined from many different analyses, providing input at different scales Q. This picture of partons (quarks and gluons) distributed in the hadrons and interacting individually relies on the asymptotic freedom and is called the parton model [15]. We can then picture the high Q^2 proton-proton reaction as a hard scale scattering between two initial state partons with large momentum transfer[3] and some additional radiation and the remnants (spectator partons, i.e. the "rest" of the hadron), where the hard scattering happens very quickly and is therefore not "seen" by the much slower interactions in the hadron. Figure 2.1 shows a schematic of this picture for W/Z production.

The process cross section for a proton-proton collision can then be written as

$$\sigma_{pp \to X} = \sum_{q_1, q_2} \int_0^1 dx_1 \int_0^1 dx_2 f_{q_1/p_1}(x_1) f_{q_2/p_2}(x_2) \hat{\sigma}_{q_1 q_2 \to X}(x_1, x_2, s) \qquad (2.21)$$

where $f_{q/p}(x)$ is the PDF for the parton q in the proton p, x is the momentum fraction carried by the parton, $\hat{\sigma}$ is the hard scattering cross section and $\sqrt{s}$ is the center-of-mass energy of the pp system. The sum runs over all parton combinations $q_1 q_2$ in the two hadrons. The fraction x of the momentum carried by the parton can take values between 0 and 1 with the parton momentum given by $p = xP$ where P is the total momentum of the incoming proton. The parton momenta are collinear to the proton's momentum in

---

[3]Most pp reactions occur at a large distance between the interacting particles, thus the momentum transfer is very small and the final state particles will have a large longitudinal but a very small transverse momentum of the order of a few MeV. This kind of events are referred to as *soft* events. On the other hand, interactions with a large transverse momentum transfer are called *hard* events.

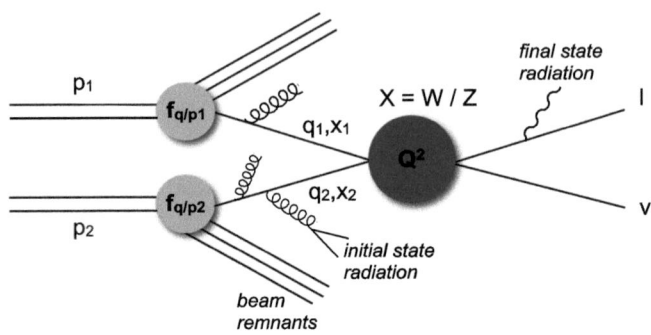

Figure 2.1: Schematic view of the W/Z production in proton-proton interactions decaying leptonicaly.

good approximation. The center-of-mass energy of the protons with momentum P_1 and P_2 is given by

$$s = (P_1 + P_2)^2 \tag{2.22}$$

and the parton-parton energy

$$Q^2 = \hat{s} = (p_1 + p_2)^2 = x_1 x_2 s. \tag{2.23}$$

The PDFs not only depend on the momentum fraction x of the parton but also on the energy scale Q. Once a PDF is known at a certain scale Q_0 it can be propagated to a scale $Q > Q_0$ with the DGLAP parton evolution equation [16, 17, 18]:

$$Q^2 \frac{\partial f_{i/p}(x, Q^2)}{\partial Q^2} = \sum_{j=g,q,\bar{q}} \int_x^1 \frac{dz}{z} \frac{\alpha_s}{2\pi} P_{i/j}(z, \alpha) f_{j/p}\left(\frac{x}{z}, Q^2\right) \tag{2.24}$$

where the sum runs over all quark flavors. The functions $P_{i/j}(z, \alpha)$ are the splitting functions which give the probability for a parton i splitting from a parton j with momentum z and are computable in perturbation theory.

The PDF sets used for LHC physics are based on fits to data from all available deep inelastic scattering (DIS) and related hard proton collisions data. Some of the most commonly used sets are those from CTEQ [19] and MSTW [20]. Figure 2.2 shows the parton distribution functions for quarks, antiquarks and gluons for two different values of Q^2 at next-to-leading order (NLO) calculation in α_s for the LHC. The sets are produced by the MSTW collaboration and are determined from global fits to data, including the newer H1 and ZEUS (HERA), and CDF and D0 (Tevatron) data. The scale dependence of the PDFs becomes visible in the two figures ($Q^2 = 10\,\text{GeV}^2$ and $Q^2 = 10^4\,\text{GeV}^2$), where the fraction of partons with small x increases with increasing Q^2 and less partons with large x are found, i.e. with rising Q^2 more and more gluons and quark-antiquark (sea quarks) are resolved.

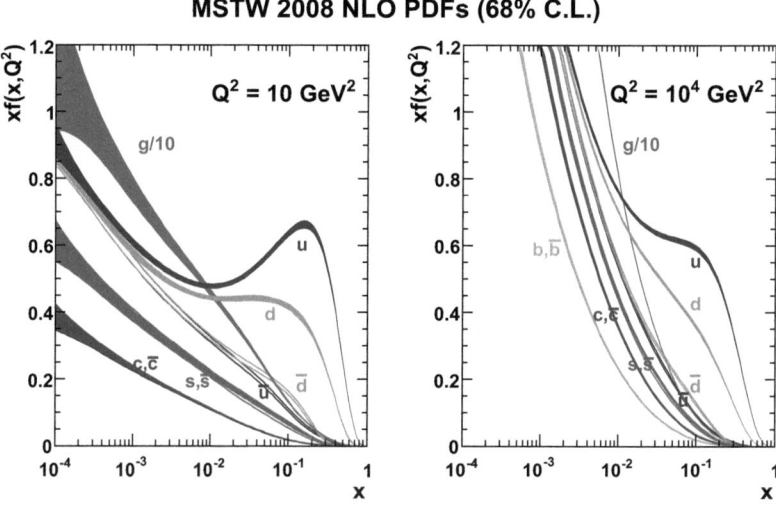

Figure 2.2: Parton distribution functions for quarks, antiquarks and gluons for the LHC at NLO for $Q^2 = 10\,\text{GeV}^2$ (left) and $Q^2 = 10^4\,\text{GeV}^2$ (right) determined from fits to data by the MSTW collaboration. The bands indicate the 68% confidence level uncertainty [20].

2.2.2 W and Z Boson Production and Decay Channels

W and Z bosons are produced in abundance at the LHC due to their large cross section. The production cross section at LHC energies is on the order of 100 nb and 30 nb and with subsequent leptonic decay around 10 nb and 1 nb per channel for W and Z, respectively [21] and after jet and b production the largest one. W and Z production is one of the best understood processes and the hard scattering cross sections have been calculated up to the next-to-next-to-leading order (NNLO) [22, 23, 24, 25] approximation.

The leading order (LO) W and Z production process is given by the annihilation of a quark and an antiquark, and depending on the flavor and sign they form a W^+, W^- or Z. The W production is dominated by $u\bar{d} \to W^+$ and $d\bar{u} \to W^-$ and the Z production by $u\bar{u} \to Z$ and $d\bar{d} \to Z$. In fig. 2.3 the diagram of the LO processes $pp \to W \to l\nu$ and $pp \to Z \to l^+l^-$ is shown.

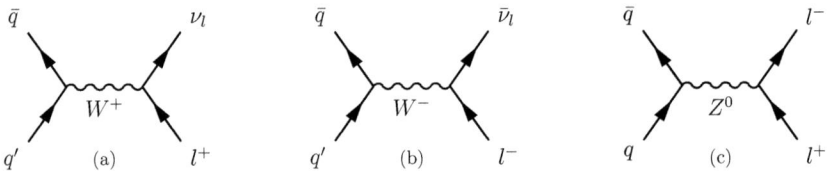

Figure 2.3: Leading order feynman diagrams for (a) W^+, (b) W^- and (c) Z^0 production with subsequent leptonic decay in proton-proton collisions.

At the LHC the Q^2 can be large and thus the fraction of gluons and sea quarks contributing in the production processes is considerable. Even though W and Z bosons are mainly produced by u and d quarks, all flavors contribute. The W and Z masses have been measured at LEP and Tevatron and a combined fit yields $M_W = 80.399 \pm 0.023$ GeV and $M_Z = 91.1876 \pm 0.0021$ [2]. For massive particle production the scale is given by the mass $M \sim Q$ and with the rapidity

$$y = \frac{1}{2} \ln \frac{x_1}{x_2} \qquad (2.25)$$

we obtain the following expression for the momentum fraction x of the two partons:

$$x_{1,2} = \frac{M}{\sqrt{s}} e^{\pm y}. \qquad (2.26)$$

Figure 2.4 shows the LHC kinematic ranges and we can see that for small values of y in the W and Z mass range ($Q \sim 10^2$ GeV) the parton's momentum fraction are of the order of 10^{-3} to 10^{-1} (e.g. for y = 0, $x_{1,2} \sim 10^{-2}$ and for y = 2, $x_1 \sim 10^{-1}$, $x_2 \sim 10^{-3}$).

The W and Z bosons are very short lived ($\sim 10^{-25}$ s) particles according to their resonance width $\Gamma = \hbar/\tau$ with $\Gamma_W = 2.085 \pm 0.042$ GeV and $\Gamma_Z = 2.4952 \pm 0.0023$ GeV, respectively [2]. They can decay into leptons or hadrons. The branching ratio (BR) for the hadronic decay is BR(W → hadrons) = $(67.60 \pm 0.27)\%$ and BR(Z → hadrons) = $(69.91 \pm 0.06)\%$ and for the leptonic decay BR(W → $l\nu$) = $(10.80 \pm 0.09)\%$ and BR(Z → l^+l^-) = $(3.3658 \pm 0.0023)\%$, with $l = e, \mu, \tau$. The invisible decay mode of the Z to two

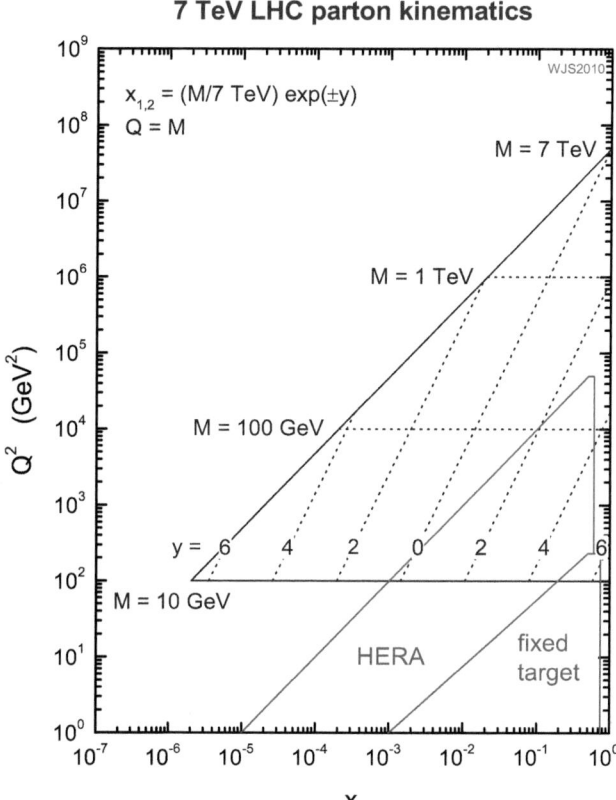

Figure 2.4: Parton momentum fraction x and energy scale Q^2 at the LHC at 7 TeV center-of-mass energy [20].

neutrinos is BR($Z \to \nu\bar{\nu}$) = (20.00 ± 0.06)%. The values are determined by the particle data group (PDG) combining measurements from different experiments and assuming lepton universality [2].

Even though the hadronic branching ratios are larger, experimentally the leptonic decays are measured. This is because it is impossible to distinguish the hadronic decays in the LHC environment, while electrons and especially muons are very easy to identify and measure through their four-vectors.

2.2.3 Jets in W and Z Events

The LO picture does not suffice to explain the processes observed in experiments. Next-to-leading and higher order calculations complement the LO hard scattering with real additional parton emissions and virtual loops. In the long-distance timescale, the description with field theory fails and phenomenological models have to be used as an aid. The picture of the $pp \to W(Z) + X \to l\nu(ll) + X$ process, where X stands for the additional final state particles can be summarized from an experimental point of view as described in the PYTHIA event generator by Sjöstrand, Mrenna and Skands [26]. The two incoming proton beams are composed of partons which can be described by the corresponding (universal) PDFs. Each of the two partons from the incoming protons which participate in the hard scattering emit QCD or QED radiation such as $q \to qg$ and $g \to q\bar{q}$ which we call initial-state radiation (ISR). The hard scattering produces the short-lived resonance W or Z which decay into the final state products, the leptons. The outgoing particles can emit additional QED final-state radiation (FSR), for instance a muon emitting a photon $\mu \to \mu\gamma$.[4] Additionally, the other partons of the two protons can interact (multi-parton interactions) leading to usually low energetic final state particles. The "rest" of the proton which does not contribute to the hard interaction is called beam remnant. It is color-connected to the hard scattering final state and carries the rest of the proton energy. Due to the color confinement the outgoing quarks and gluons fragment and form new colorless objects, collimated hadrons which we call jets. This process is called hadronization or fragmentation.

Due to the ISR added to the LO picture the corresponding parton and hence the W (Z) boson have obtained a momentum component transverse to the beam direction. The radiated parton hadronizes and a jet is formed additionally to the W (Z) bosons, with compensating transverse momentum. Figure 2.5 shows the Feynman diagram for the first order W + jet production where the produced jet originates from a gluon (a) or a quark (b) which radiates before interacting with the parton from the other proton.

The QCD emissions become more and more important with increasing energy and at the LHC, W(Z) events with one and more jets are produced. As was mentioned before, the corrections can in principle be calculated order by order in perturbation theory. For instance the NLO calculations include the real corrections producing the hard object (the jet) in the final state and also all the virtual corrections. Thus, with increasing order the number of diagrams which have to be calculated grows and the calculation becomes very

---

[4]In general FSR includes gluon or quark emission but as leptons do not possess color, in the leptonic decay only photons can be emitted.

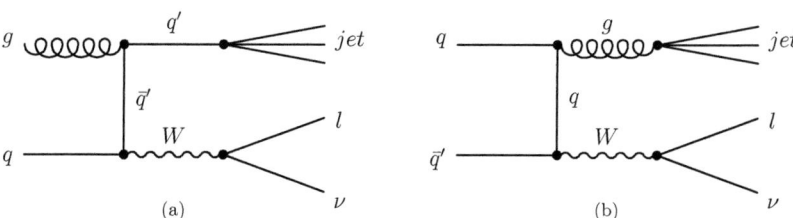

Figure 2.5: Feynman diagrams for the first order W + jet production where the produced jet originates from a gluon (a) or a quark (b) which radiates before interacting with the parton from the other proton

tedious. A good approach to estimate the corrections rather than the exact calculation is the parton showering (see e.g. [27, 28, 29] and references therein). Here, the complete final state e.g. a multijet topology can be described with any number of involved partons by only using approximations with simplified kinematics, and interference and helicity structure instead of the full expressions. The approximation delivers a probability splitting function $P_{a \to bc}(z)$ for a parton a with a certain momentum p to branch into two daughter partons b and c with momentum fraction z and $1-z$, respectively. The partons b and c can then branch further and iteratively the daughters of them. The evolution of the shower can be done in different ways, depending on the variable used for the time ordering. The time ordering of the evolution can be described by the virtuality scale Q^2 of the parton. For FSR the Q^2 decreases gradually until reaching a minimal value Q_0 where the evolution is stopped. Similarly, for ISR the Q^2 is set to rise gradually up to a certain cut-off value Q_{max} where the shower is matched to the hard interaction. Different definitions of the evolution variable Q^2 exist, the ones used in PYTHIA being *virtuality-* or *mass-ordered* and in most recent versions p_T-*ordered* shower algorithms. In the mass-ordered showers Q^2 is defined as $Q^2 = m^2 = E^2 - \mathbf{p}^2 \geq 0$ for the time-like FS showers and $Q^2 = -m^2 \leq 0$ for the space-like IS showers. The p_T−ordered showers, as the name says are defined by $Q^2 = p_T^2 = z(1-z)m^2$ (FSR) and $Q^2 = p_T^2 = -(1-z)m^2$ (ISR), with p_T defined as the transverse momentum. The p_T−ordered showers are the more recent implementations and have some advantages over the older virtuality-ordered showers, particularly concerning multi-parton interactions which are interleaved with the ISR in the new algorithms. Finally, after the showering evolution the hadronization is applied.

The hadronization process is not yet understood completely. Models exist which allow for precise predictions and description of experimental data. There are three different basic types: the string fragmentation [30, 31], the cluster fragmentation [32], and the independent fragmentation. The first two are the most widely used at the LHC and implemented in the PYTHIA (mainly used in this theses) and HERWIG event generators, respectively. Maybe most important for this thesis is that the different model implementations of fluctuations in the hadronization process will results in distinct probabilities of finding regions in the detector with a down fluctuation. This, as we will see in section 6.1, will have an influence on the studied distributions and especially on the event signature for diffractive events. The basic idea of the string model is that as $q\bar{q}$-pairs move apart, the color field between them forms a colour flux tube (the strings) which is being stretched. The potential energy stored rises with the stretching until it leads to the creation of another $q\bar{q}$-pair, allowing the string to break and forming two color-singlets. If the invariant mass of one of the two new pairs is large enough, the process continues until only on-mass-shell hadrons remain in the jet. These may decay further forming the final stable particles measured in the detector.

2.2.4 Underlying Event

Additionally to the hard interaction the beam remnants leave their trace in the detector. They carry a large fraction of the total proton energy and are color-connected to the primary interaction, thus they are part of the same fragmentation system. The description of the remnant gets complicated by the fact that the remnant partons can subdivide into subsystems with different possibilities of energy sharing and which are also color-connected to the hard interaction. Additionally, the probability for other partons than the ones forming the hard interaction to interact is non-negligible. Most of the time these interactions will be soft but in some events hard or semi-hard scattering may occur. These multi-parton interactions (MPI) manifest in the underlying event structure of the event and result in an increased total multiplicity. We define the underlying event (UE) as everything else except the hard scattering, i.e. the low energy hadrons from the parton showers and the (soft) MPI. Moreover, at the LHC where the protons cross in entire bunches rather than one single proton, the probability for more than one proton colliding event increases with the luminosity. These additional events are called pileup events (PU) and again have an influence on the resulting measurements in the detector.

2.2.5 Diffractive Processes

The total cross section of the proton-proton interaction at the LHC can be divided in elastic and non-elastic processes, where the largest contribution comes from the latter. The cross section was recently measured with the TOTEM (TOTal cross section, Elastic scattering and diffraction dissociation Measurement at the LHC [33]) experiment at the LHC at a centre-of-mass energy of $\sqrt{s} = 7\,\text{TeV}$ and is $(98.3 \pm 0.2^{\text{stat}} \pm 2.8^{\text{syst}})\,\text{mb}$. The elastic scattering cross section was measured to be $(24.8 \pm 0.2^{\text{stat}} \pm 1.2^{\text{syst}})\,\text{mb}$ [34]. From the total and elastic pp cross section measurements, an inelastic pp cross section of $(73.5 \pm 0.6^{\text{stat}\,+1.8\text{syst}}_{\phantom{\text{stat}\,}-1.3})\,\text{mb}$ was obtained, consistent, within the errors, with the CMS measurement of $(68.0 \pm 2.0^{\text{syst}} \pm 2.4^{\text{lum}} \pm 4^{\text{ext}})\,\text{mb}$ obtained in the same year [35].

A considerable fraction of the total pp cross section is expected to arise from diffractive reactions, which can be elastic or inelastic interactions. Overall, diffractive final states, including elastic scattering, are estimated to be about 50% of all final states at the LHC [36]. Most of the physics under investigation at the LHC happens in the non-diffractive inelastic region. However, a large part of the inelastic scattering cross section corresponds to the several diffractive production types, where the inclusive single diffractive (SD) events, $pp \rightarrow pX$ are expected to amount approximately 15% of the total cross section. Figure 2.6 depicts the total pp cross section including the different final states of diffractive events.

Single Diffractive Events

Single diffractive events are processes where one of the colliding protons emerges intact from the interaction, having lost only a few percent of its energy. Such events may be ascribed to the exchange of a colorless object with the vacuum quantum numbers (often called Pomeron exchange). After the exchange, the proton can either remain intact or dissociate, the latter being called diffractive dissociation. Diffractive dissociation leads to the absence of hadron production over a wide region of rapidity adjacent to the outgoing proton direction. Experimentally, these large rapidity gaps will appear as regions of pseudorapidity devoid of detected particles. In fig. 2.7 the detector is sketched for several diffractive event signatures, picturing the empty regions where no particles are deposited.

Soft-diffractive events can be described in the framework of Regge theory (see e.g. [37]). Hard-diffractive events, with the production of jets, heavy flavors, or W/Z bosons, have been observed at the SPS, HERA, and the Tevatron [38, 39, 40, 41, 42, 43]. From electron-

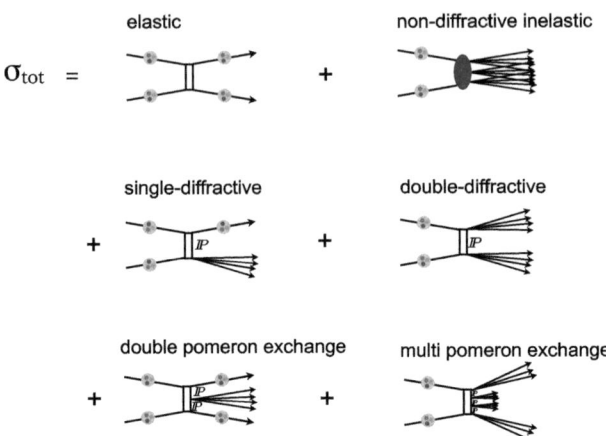

Figure 2.6: Diagrams of the various components of the total proton-proton cross section at the LHC.

proton hard-diffractive interactions it is known that the pomeron can be described by partons consisting mainly of gluons. Also, a factorization theorem has been proven [44], allowing the introduction of diffractive parton distribution functions (dPDFs). In hadron-hadron diffractive interactions, factorization is however broken by soft multi-parton interactions [45, 46], which fill the large rapidity gap and reduce the observed yields of hard-diffractive events. Such interactions are not yet simulated in the currently available Monte Carlo (MC) generators, and the reduction in the large rapidity gap hard-diffraction cross section is quantified by a suppression factor, the so-called rapidity gap survival probability. At the Tevatron, the observed hard-diffractive yields relative to the corresponding inclusive processes are approximately 1%. A recent study by CDF [47] indicates that the fractions of W and Z bosons produced diffractively are $(1.00 \pm 0.11)\%$ and $(0.88 \pm 0.22)\%$, respectively.

2.3 Monte Carlo Event Generators

As stated above, the description of the physics in proton-proton collision events is very complex and relies on many approximations and models. Yet, it is very important to be able to simulate these events, starting from the requirements to build a detector, to study

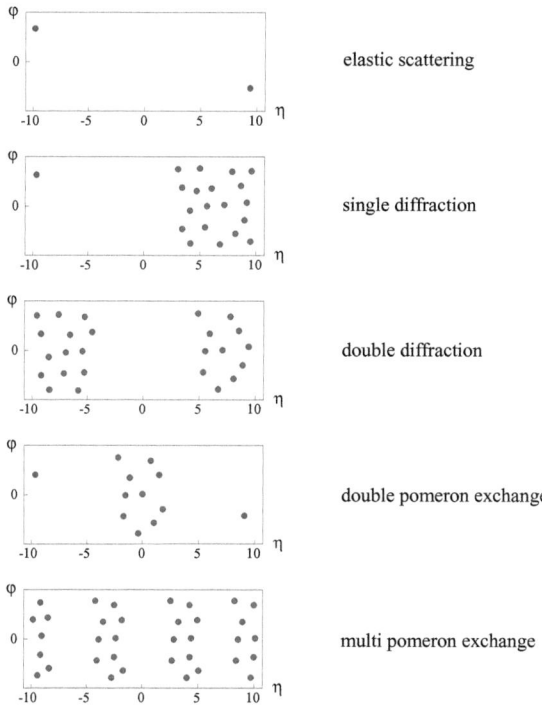

Figure 2.7: Sketch of the $\eta - \phi$ detector view for the various diffractive processes in proton-proton scattering at the LHC.

and predict event topologies for expected new physics, to the understanding of detector effects and the final comparison with the real data. The aim of event generators is to produce events which describe the real events as accurately as possible. The task has to be split up in several steps, close to the steps elaborated in the sections before. First, the perturbatively computed hard cross sections are used as input. Then, a parton shower algorithm is applied to include the corrections to the calculated n^{th}-order cross sections. The multiparton interactions have to be taken into account, after which the fragmentation procedure follows, and the subsequent decay to the final state particles. The quantum mechanical fluctuation of real data is simulated by Monte Carlo techniques taking into account the corresponding probability functions. The final prediction and comparison to data is done by feeding the output to the detector simulation based on GEANT4 [48] and the entire event reconstruction chain. We will use the naming *generator-* or *particle-level* and *detector-level* in the course of this thesis, meaning the level before and after the detector simulation step, respectively. In this thesis two different event generators are used, the widely used PYTHIA event generator and the MadGraph program interfaced with PYTHIA.

In PYTHIA the W and Z production cross sections are calculated at LO and convoluted with the CTEQ 6.6 (default) parton distribution functions (PDFs) [49]. Effects from higher-order QCD corrections are approximated with parton showers from the initial and final-state partons and for hadronization the string fragmentation model is used.

The currently available models of multi-parton interactions have mainly been tuned to minimum-bias data[5] and to final states including jets with large transverse momentum, using the observed central charged-particle multiplicities and the transverse momentum spectra of hadrons that are not associated with the hard jets. Models for a detailed simulation of multi-parton interactions are under rapid development and new features, including diffractive components in the energy flow, are being extensively investigated.

The simulation of non-diffractive processes, including pileup, used in this work, is done with the PYTHIA6.420 and PYTHIA8.145 event generators [26, 50] with different tunes [51] for the underlying event structure and the multiparton interactions. The tunes developed before data from the LHC are PYTHIA6 D6T [52], Pro-Q20 [53], Pro-PT0 [53], and P0 [54]. The newer tunes PYTHIA6 Z2 [55] and the PYTHIA8 2C [56] include already some information from the LHC data. The older D6T and Pro-Q20 tunes are associated with virtuality-ordered showers, while the newer ones, P0, Pro-PT0, Z2, and 2C, are associated with p_T-ordered showers. The newer p_T-ordered showers have some improved implementations of the multiparton interactions which are interleaved with the ISR and thus a better agreement is expected with the newer tunes when comparing to data distributions (cf. chapter 7). As shown in the recent CMS measurement of the underlying event structure [57], the Z2 tune provides a reasonably good description of the data in minimum-bias events at central rapidities.

Diffractive W and Z production are simulated with the POMPYT 2.6.1 [58, 59, 60] event generator. The hard processes responsible for the production of W and Z bosons are identical to those in non-diffractive models. For the simulation of diffractive processes, the dPDFs (fit B) measured by the H1 experiment at HERA are used [41, 61]. This generator does not simulate multiparton interactions or the ensuing rapidity gap survival probability.

The newer data from the 2011 run is compared to the generated events with the MadGraph

---

[5]Events with minimal trigger conditions, dominated by soft events.

program [62] interfaced to PYTHIA. MadGraph is a matrix element generator which automatically generates the amplitudes for the relevant subprocesses, given a specific SM process, and passes them to the multi-purpose tree level event generator MadEvent [63]. The generated event information is then passed to PYTHIA, where the showering and hadronization is done. For the PYTHIA part, virtuality ordered showers are used and the PDF set CTEQ6L1. All relevant generator parameters can be found at the MadGraph website [64]. For the background processes the PYTHIA event generator was used, except for $t\bar{t}$. In table 2.2 the MC samples for signal and background used in this thesis are listed. At generator level, the cross sections are calculated at LO. To weight the different datasets according to the data luminosity, NNLO, NLO and NNLL cross sections were used and are indicated in the table.

Table 2.2: Monte Carlo datasets for signal and background.

Process	Generator	PDF	cross section [pb] x BR
W $\to \mu\nu$	PYTHIA6(8), several tunes	CTEQ6.6	7899 (LO)
Z $\to \mu\mu$	PYTHIA6(8), several tunes	CTEQ6.6	1300 (LO)
WX $\to$ lνX	MadGraph/Pythia	CTEQ6L1	31314 (NNLO)
ZX $\to$ llX	MadGraph/Pythia	CTEQ6L1	3048 (NNLO)
$t\bar{t}$	MadGraph/Pythia	CTEQ6L1	158 (NNLL)
QCD	PYTHIA6, Z2 tune	CTEQ6LL	84679.3 (LO)
WW	PYTHIA6, Z2 tune	CTEQ6LL	47 (NLO)
WZ	PYTHIA6, Z2 tune	CTEQ6LL	18.3 (NLO)
ZZ	PYTHIA6, Z2 tune	CTEQ6LL	7.67 (NLO)

Chapter 3

The CMS Experiment at the LHC

The CMS experiment is designed to study a large variety of physics at the TeV scale. The technological challenges for such high-energy measurements are manifold. High precision is required in space and time for the detection of the particles; fast electronic has to cope with huge data amounts at high rates; the detector components have to be resistant to large radiation exposure; it should be as hermetic as possible in order not to loose particles in detector gaps. These constraints have been addressed in a layered, cylindrical detector construction. In this chapter, a detailed description of the CMS detector and its subcomponents is given. Beforehand, the LHC is introduced shortly, and the chapter is concluded with a brief overview of the two data taking periods relevant to this thesis and a selection of CMS measurements related to W and Z physics.

3.1 The Large Hadron Collider

The Large Hadron Collider (LHC) [65] is a particle accelerator situated at CERN near the Swiss-French border. It is placed in a tunnel excavated between 50-175 m under ground with a circumference of 26.7 km. It also hosted the preceding accelerator at CERN, the Large Electron-Positron collider (LEP). It is constructed to collide two counter rotating beams of either protons or lead ions. In the proton modus, it was designed to run at a center-of-mass energy of $\sqrt{s} = 14\,\text{TeV}$ and a nominal instantaneous luminosity of $\mathcal{L} = 10^{34}\,\text{cm}^{-2}\text{s}^{-1}$. This large increase in both energy and luminosity compared to previous collider experiments is aimed to explore the vast possibilities of physics beyond the electroweak scale and test the limits of the standard model. The number of expected events for a certain process, given by

$$N = \mathcal{L} \cdot \sigma$$

depends on these values, which is the reason to keep them as high as possible. The dependance of the cross section on the center-of-mass energy for different physics processes is shown in fig. 3.1.

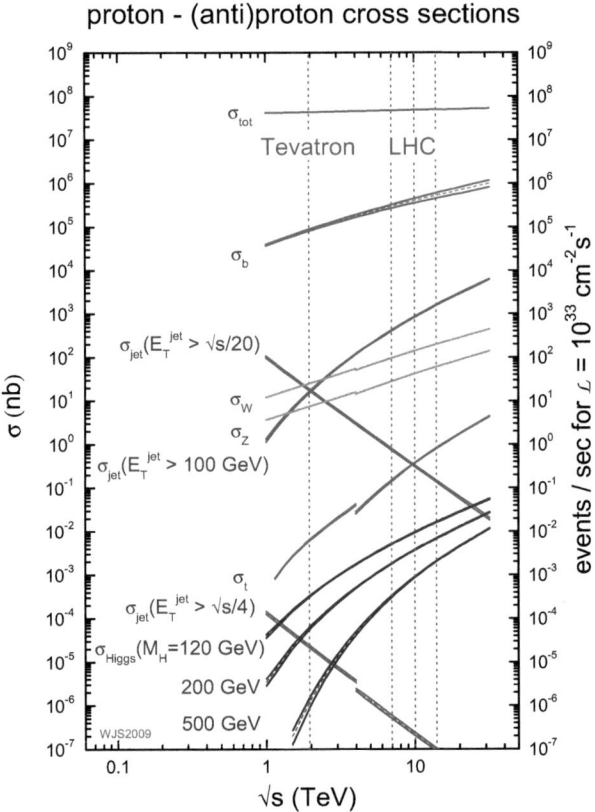

Figure 3.1: Cross sections for different processes as a function of the center-of-mass energy [66].

The luminosity is given by
$$\mathcal{L} = f \cdot n_b \frac{N_1 N_2}{A}$$
where f is the revolution frequency, n_b is the number of proton bunches of which the beam is made of, N_1 and N_2 are the number of particles in each bunch and A is the effective transverse area of the beam given by $4\pi\sigma_x\sigma_y$, and $\sigma_{x/y}$ the transverse beam size at the interaction point. The design parameters of the machine are 2808 bunches per beam, each bunch containing 10^{11} protons and separated by 25 ns (equivalent to 7.5 m). The beam size is minimal at the interaction point and is around 7.5 cm in longitudinal and 16 μ in transverse direction. At nominal luminosity, this results on average in 20 collisions per bunch crossing and 800 million collisions per second in each interaction point.

In late 2009 a commissioning run was done at $\sqrt{s} = 900$ GeV and $\sqrt{s} = 2.36$ TeV. The first two full years of operation which started on the 30th March 2010, the LHC ran at $\sqrt{s} = 7$ TeV, increasing it to $\sqrt{s} = 8$ TeV beginning of 2012. So far, the shortest bunch crossing achieved is 50 ns. The maximum peak luminosity reached in 2011 amounted 3.6x10^{33} cm^{-2}s^{-1} and has already been exceeded in the 8 TeV running period.

The acceleration of the proton beams is increased in several steps in the CERN acceleration complex. Figure 3.2 shows a schematic overview of the several accelerators including the LHC. Protons obtained from hydrogen atoms are consequently injected from LINAC2, to the PS Booster, the Proton Synchrotron (PS), the Super Proton Synchrotron (SPS) and finally to the LHC. Gradually they acquire more energy and after circulating for about 20 min they eventually reach the maximum velocity.

The LHC machine is composed of vacuum tubes where the particles circulate in ultrahigh vacuum. Superconducting dipole magnets with a field up to 8.3 T are used to keep the particles in their orbit. Quadrupole and sextupole magnets focus the beam in transverse and longitudinal direction, especially in front of the interaction point. Acceleration cavities (radio frequency resonators) accelerate the particles and in the next stage keep them at constant energy to compensate for energy loss.

Along the LHC four detectors are located around the collision points (fig. 3.2). Two of them, CMS [68] and ATLAS [69] are general purpose detectors while the LHCb [70] experiment is designed to study B-physics and the ALICE [71] detector is dedicated to physics of heavy ions.

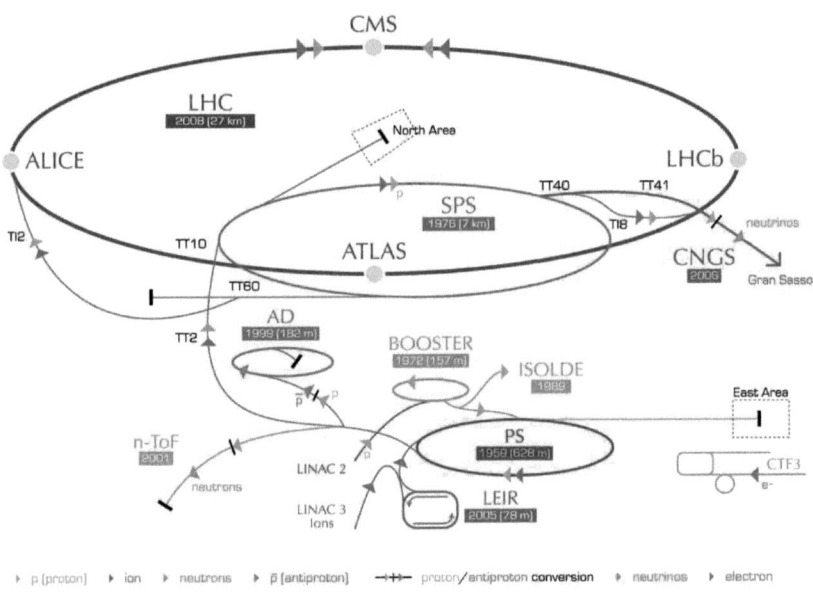

Figure 3.2: Overview of the CERN acceleration complex. The location of the four experiments along LHC are indicated [67].

3.2 The CMS Detector

3.2.1 Overview

The Compact Muon Solenoid (CMS) is a general purpose experiment composed of several layered detector components. It is situated 100 m underground at one of the LHC collision points in France. It is 21.6 m long and 15 m in diameter and weights 12500 t. The detector has to meet several design objectives in order to fulfill the manifold physics challenges. The main challenges are the huge background in pp collisions compared to the production rates of the processes being investigated at the LHC, the fast rate at which the collisions occur, and the large radiation exposure of the material from the huge particle flux. For these reasons the detector has to be able to measure the diverse signatures of the particles with an excellent identification efficiency and momentum resolution, especially muons and electrons which are experimentally favored over hadronic final states and thus essential to many of the searches, together with an excellent timing resolution, fast response and radiation hardness. To meet these requirements the detector has been constructed in several layers of sub-components, each of them providing complementary information. They are arranged in layers around the beam pipe, forming a cylindrical shaped detector. From inside to outside, CMS is composed of the tracking system delivering precise charged particle momentum resolution and secondary vertex reconstruction for b-jets, τ and pileup identification; the electromagnetic and hadronic calorimeters for identification of photons, electrons and hadrons and for a calorimetric energy measurement; the magnet to bend the particles; and the outermost layer, the muon spectrometers for an excellent muon identification and momentum resolution interleaved with iron return yokes to contain the magnetic field. Additionally, the hermeticity of the detector is indispensable for missing energy and jet energy resolution, for cross section and diffractive measurements. The cylindrical shape of the detector is formed by a *barrel* and two *endcaps* closing the front and back of the detector, with only two small gaps for the beam pipe. Figure 3.3 shows a drawing of the CMS detector with the several sub-components, described in more detail in the following. Figure 3.4 shows a sketch of a slice of CMS transverse to the beam axis. The trajectories of the different particle classes and their signature in the several detector layers are depicted.

The coordinate system of CMS is chosen such that the point of origin is the collision point, the x axis points south to the center of the ring, the y axis vertically upwards and the z axis correspondingly to the west tangential to the beam line forming a right handed system. Spherical, some times cylindrical coordinates are used with the azimuthal angle

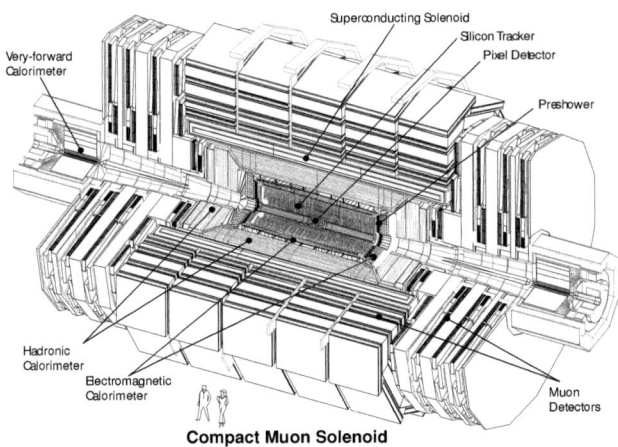

Figure 3.3: Schematic view of the Compact Muon Solenoid with the several detector components indicated [72].

ϕ measured in the xy-plane starting from the x axis. The polar coordinate θ is measured in the rz-plane starting from the z axis. Due to it's Lorentz-invariance the pseudorapidity $\eta = -\ln\left(\tan(\frac{\theta}{2})\right)$ is usually taken as reference instead of θ. The xy-plane is from now on called transverse plane and 'T' refers to the value measured in that plane, e.g. transverse momentum is defined as $p_T = \sqrt{p_x^2 + p_y^2}$ or transverse energy as $E_T = E \cdot \sin\theta$ (depending on the quantity which is measured by the detector).

In the following, each subcomponent is briefly described. More details about CMS and the subdetectors can be found in [68].

3.2.2 Magnet

The 'Solenoid' in 'CMS' refers to the superconducting coil magnet located between the muon system and the hadronic calorimeter. Its purpose is to bend the particle trajectories according to $R \sim \frac{p}{0.3 \cdot B}$, [1] where R is the radius, q the charge and B the magnetic field. Thus, through the radius R and the sign, the momentum of charged particles and their charge can be obtained. The momentum resolution which can be obtained depends on

---

[1]From the equation of motion for a charged particle on a circular path resulting from the Lorentz force of the magnetic field.

The CMS Experiment at the LHC

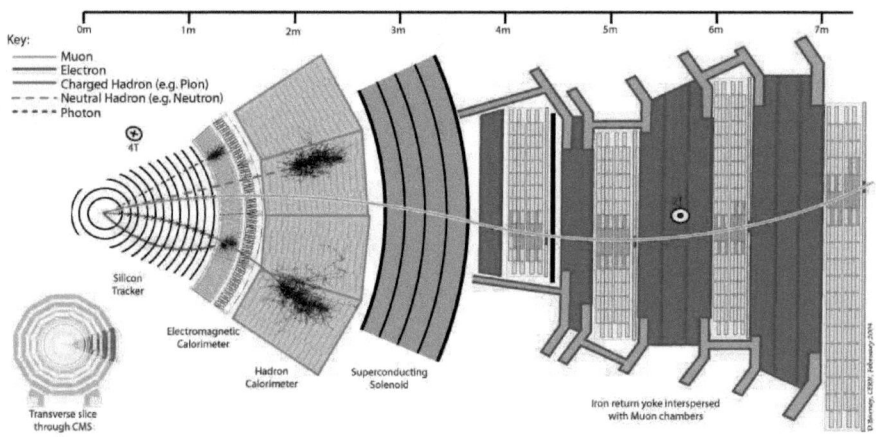

Figure 3.4: Slice of CMS transverse to the beam axis with the detector components and the trajectories of the different particle classes through the detector [73].

the magnetic field B but also on the magnet length L along which the particle travels:

$$\frac{\Delta p}{p} \sim \frac{p\Delta s}{qL^2 B}$$

with s the sagitta of the bending path. This constrains the choice of the magnet system along with the rest of the detector design, which has to 'fit' to the magnet. A solenoidal magnet composed of four layers winding of NbTi was chosen with a uniform axial field of 3.8 T and an outer diameter of 6.5 m, inner diameter of 5.9 m and an axial length of 12.9 m. The drawback of this choice is the need of a return yoke to contain the magnetic flux which leads to multiple scattering and therefore degrades the resolution (for low momentum) and moreover the magnetic stray field has to be properly understood and described.

3.2.3 Inner Tracking System

The tracking system is the innermost subdetector of CMS and measures the tracks charged particles. Its aim is to provide a robust, efficient and precise reconstruction of the charge, position and momentum of the particles, secondary vertices and the impact parameter. Secondary vertices are the decay vertices of long lived (order of few hundred μm) particles like heavy quarks or taus and are needed to identify them. The impact parameter is the transverse distance from the reconstructed track to the interaction point and can be used e.g. as a measure of the primary vertex quality or to discriminate cosmic muons. Moreover, the tracking system together with the adjoining calorimeters are used for particle identification.

Due to the high particle flux close to the interaction point, the main challenge in the tracking system consists in providing at the same time a high granularity and a fast response with the least material budget possible to limit multiple scattering, bremsstrahlung, photon conversion and nuclear interactions, and maximize the radiation resistance of the detector. In CMS, an all-silicon tracking system is used, composed of an inner pixel detector and a micro-strip detector surrounding it. The cylindrical tracker has a length of 5.8 m and a diameter of 2.5 m with a total active area of 200 m^2. It is composed of a barrel and two endcaps covering a pseudorapidity range up to 2.5. Figure 3.5 shows a schematic view of the tracker layout in the $r - z$ plane with the several detector components.

The pixel detector covers a radius from 4.4 cm to 10.2 cm. It is composed of 3 barrel layers and two endcap disks on each side. At a radius of 4 cm and design luminosity the hit rate density amounts 1 MHz/mm^2. To obtain an occupancy of the order of 10^{-4} in the detection unit, a pixel layout is used with a size of $100 \times 150\,\mu m^2$.

With increasing radius the flux decreases and the pixels can be replaced by strips. The outer part of the tracker is made of 10 barrel strip layers covering radii from 0.2 m to 1.1 m, and 3 inner and 9 outer endcap disks on each side. The strips in the inner part (20 cm$< r <$55 cm) have a typical size of 10 cm$\times$80 μm with an occupancy of 2-3% and in the outer part (55 cm$< r <$110 cm), sizes up to 25 cm$\times$180 μm leading to an occupancy of about 1%.

Combining the pixel and the strip detector, an overall tracker momentum resolution ranging from $\Delta p_T/p_T^2 = 0.015\%$ for $|\eta| < 1.6$ up to $\Delta p_T/p_T^2 = 0.06\%$ for $|\eta| = 2.5$ and an efficiency of more than 98% for $|\eta| < 2.5$ is obtained.

3.2.4 Electromagnetic Calorimeter

The Electromagnetic Calorimeter (ECAL) is a homogeneous crystal-based scintillating calorimeter which surrounds the tracking system and measures and absorbs the energy of electrons and photons and partly the one of hadrons. As was the case before, it has to provide fine granularity, fast response, and high radiation resistance. The best energy resolution for electrons and photons is given by crystal calorimeters where most of the

The CMS Experiment at the LHC 33

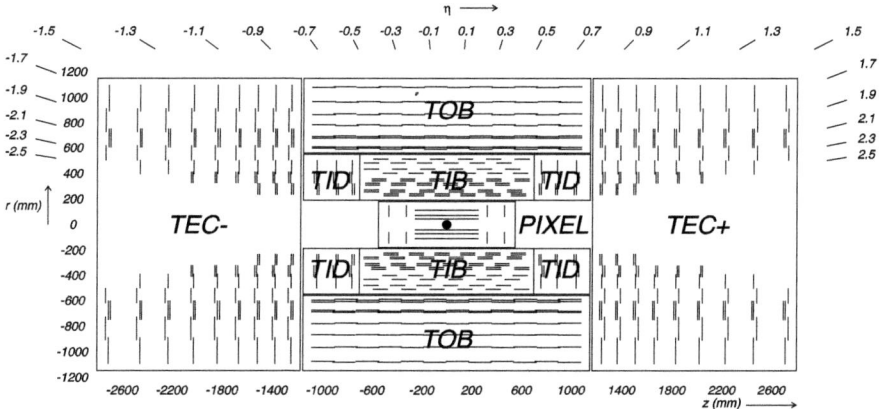

Figure 3.5: The CMS tracking system composed of an inner silicon pixel and an outer silicon strip detector in the $r - z$ plane. The strip tracker is composed of 6 outer barrel layers (TOB), 4 inner barrel layers (TIB) and on each side 3 inner disks (TID) and 9 outer endcap disks (TEC) [68].

particles energy is deposited in the active volume. The crystals in the ECAL are made of lead-tungstate (PbWO$_4$) which have a fast scintillation decay time, short showers and a good lateral containment and good radiation hardness. With 80% of the scintillation light emitted in 25 ns, the light decay time is of the same order as the bunch crossing frequency at the LHC. The radiation length is such that a crystal length of 23 cm, corresponding to 26 radiation lengths is required to contain the shower at central pseudorapidity. In the barrel the crystals have a length of 23 cm and a front face of 22×22 mm^2, in the endcaps 22 cm in length and a front face of 28,6×28,6 mm^2. To increase the spacial precision a preshower is installed in front of the endcaps. It is a sampling calorimeter made of two lead layers as absorber interleaved with silicon strip detectors for the measurement of the charged particles produced in the shower. It covers the range $1.65 < |\eta| < 2.6$ and has a thickness of 3 radiation length. Additionally it allows to discriminate single high-energetic photons from neutral pions decaying to photon pairs.

The ECAL barrel ranges from r = 1.24 to r = 1.86 and covers the pseudorapidity range $|\eta| < 1.5$. It is composed by 36 supermodules, each containing 4 modules. The modules assemble from 400 to 500 crystals, depending on the location. The ECAL endcaps are made of two Dees located at 3.2 m from the interaction point and cover the pseudorapidity range $1.5 < |\eta| < 3.0$. The crystals are combined to 5×5 crystals forming a supercrystal. Figure 3.6 shows an overview of the layout of the ECAL with its substructures in the barrel and endcaps.

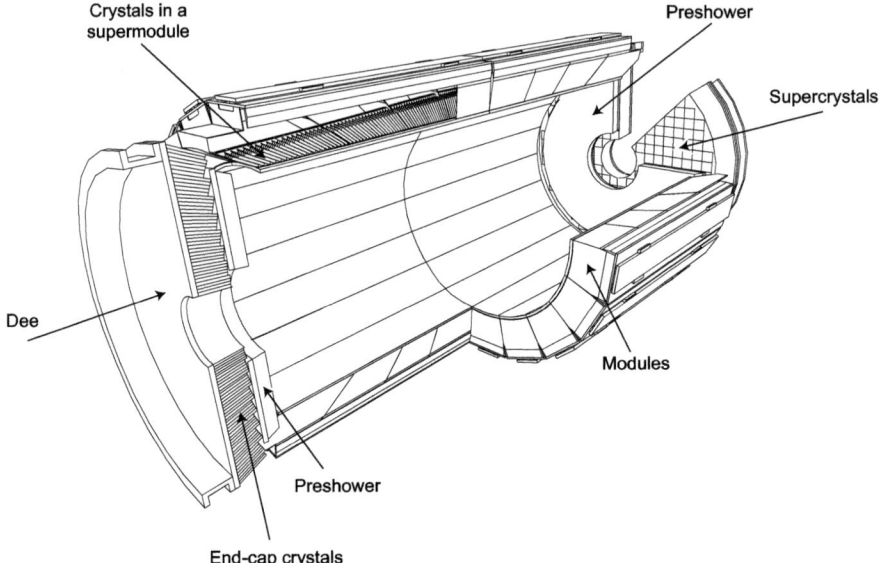

Figure 3.6: Layout of the CMS ECAL detector with the different components in barrel and endcaps [68].

A disadvantage of the lead-tungstate crystals is the low yield of the scintillation light. To compensate for this, a large amplification is needed. In the barrel, avalanche photodiodes are glued to the back of the crystals while in the endcaps vacuum phototriodes are used to amplify and readout the electrical signal.

Shower leakage from the rear of the calorimeter becomes important below 500 GeV. At these energies the resolution can be described by the following parametrization:

$$\left(\frac{\Delta E}{E}\right)^2 = \left(S/\sqrt{E}\right)^2 + (N/E)^2 + C^2$$

where S is the stochastic term, N the noise term and C a constant term. The stochastic term accounts for fluctuations in the shower containment and photostatistics. The noise term includes electronic readout noise and pileup energy. The three free parameters where determined from the measured ECAL resolution in a test beam taken in 2004 and are $S = 0.028\,\text{GeV}^{1/2}$, $N = 0.12\,\text{GeV}$ and $C = 0.003$ which agree perfectly with the design values. Figure 3.7 shows the energy resolution as a function of the reconstructed energy in a 3×3 crystal matrix as measured from the 2004 test beam. The stochastic, the noise and the constant term are indicated.

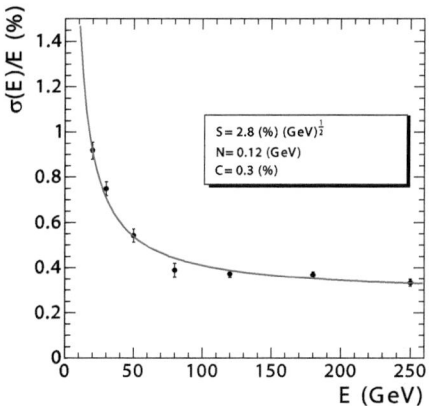

Figure 3.7: ECAL energy resolution as a function of the reconstructed energy in a 3×3 crystal matrix as measured from a test beam taken in 2004 [68].

3.2.5 Hadron Calorimeter

The Hadronic Calorimeter (HCAL) complements the ECAL by measuring the energy of hadrons, absorbing their energy through induced particle shower. Only hadrons will leave a large energy signature in the HCAL, since electrons and photons have a shorter absorption length and are already absorbed in the ECAL and Muons are Minimal Ionizing Particles (MIP). The HCAL is a sampling calorimeter with layers of absorbing brass plates alternated with plastic scintillator tiles arranged in trays. The hadrons interact strongly with the nucleus of the brass producing other hadrons with lower energy. Successively these secondary hadrons interact with the absorber and a shower is produced. In the scintillating layers the energy is re-emitted as light, sent through wavelength shifting fibers and then through transparent fibers to be finally read-out by hybrid photodetectors. To absorb the energy shower of hadrons which can be at the TeV scale approximately 10 nuclear interaction length are needed. Since the geometry is constrained by the presence of the magnet and the ECAL to a radius between 1.77 m and 2.95 m, a complementary outer calorimeter (HO) is placed between the magnet and the muon detectors.

Like the ECAL, the HCAL is divided in barrel (HB and HO) covering $|\eta| < 1.3$ and endcaps (HE) at $1.3 < |\eta| < 3$. Additionally two hadron forward detectors (HF) cover the range $2.9 < |\eta| < 5.2$. The HF modules are located at $z = \pm 11.2$ m (front face) from the interaction point and cover 40% of the available phase space of CMS. They are especially important to improve the missing energy measurements and for forward physics related analysis like diffractive processes. The HF detectors are based on radiation hard quartz fibers running parallel to the beam line embedded in a steel absorber structure. The fibers collect the Cherenkov light emitted by the showers. Bundled fibers are then routed to photodetectors located at the outer parts of the calorimeter, where the radiation doses are lower. Due to the forward location of the HF it is exposed to a very large radiation. Roughly a factor 8 more energy (760 GeV on average) is deposited per proton-proton interaction in the HF compared to the rest of the detector. This harsh environment was the main constraint and the reason for choosing quartz fibers as active material. The modules are arranged in 18 wedges (each covering 20°) in the $r - \phi$ plane forming 13 concentrical rings. The sensitive area ranges from r = 12,5 cm ($|\eta| = 5.205$ at $|z| = 11.2$ m) to r = 130 cm ($|\eta| = 2.866$ at $|z| = 11.2$ m) and each ring covers a pseudorapidity range of 0.175, except ring 1 and 13 (0,111 and 0.300, respectively). One tower is formed by bundled fibers of $\Delta\eta \times \Delta\phi = 0.175 \times 0.175$ (10°). Half of the fibers run over the length of the absorber of 165 cm (10 nuclear interaction length) and the other half starts at a depth of 22 cm, which is enough to discriminate hadrons from electrons or photons which deposit a large fraction of the energy in the first 22 cm. Figure 3.8 shows a schematic view of the transverse segmentation of the towers and the tower sizes, number of fibers and bundle sizes for each tower.

3.2.6 Muon System

The muon system at CMS allows to identify and measure muons with great precision. Their transverse momentum is measured first in the inner tracking system. Muons in the GeV range loose energy mainly by ionization[2]. As their lifetime is large enough to cross the entire detector and they only leave minimal ionizing energy in the calorimeters they reach the outer tracking system, the muon system with almost full energy. All other particles except the neutrinos are absorbed in the calorimeters. Thus, the identification of muons is in the ideal case unambiguous and with the momentum measured twice leading to an excellent resolution.

A schematic view of one quarter of the muon system in the longitudinal plane is shown in fig. 3.9. It consists of the barrel covering the range $|\eta| < 1.2$ and two endcaps at $0.9 <$

---
[2]Bremsstrahlung becomes relevant only above several hundred GeV

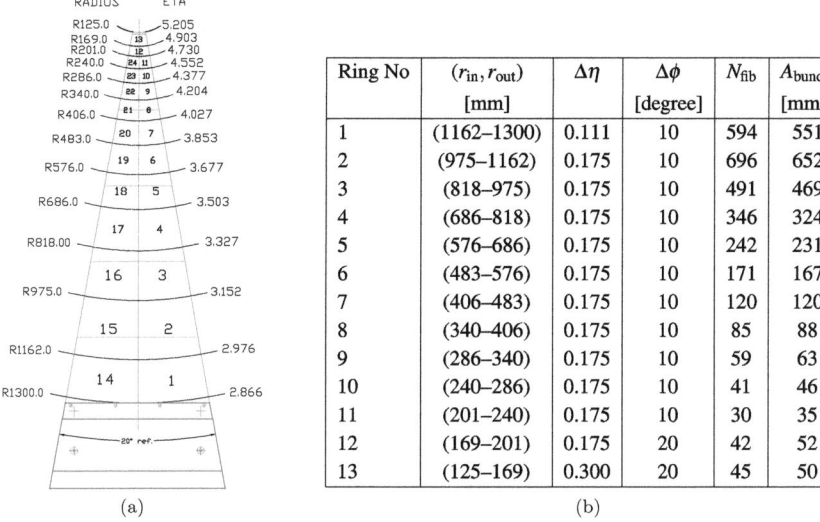

Figure 3.8: (a) Schematic view of the transverse segmentation of the HF towers and (b) tower sizes, number of fibers and bundle sizes for each tower [68].

$|\eta| < 2.4$. The barrel is composed of five iron wheels placed concentrically around the beam axis with 4 staggered detector layers (MB1-MB4) alternated with magnet return yokes distributed. The wheels are divided in 12 sectors along the ϕ direction, each section covering 30°. The endcaps consist of four disks (ME1-ME4) each, where each disk is composed of 2 concentric rings. Three different gaseous detector types are used in the muon system, depending on the particle flux and the residual magnetic field in the specific region. In the barrel, where the particle flux and the magnetic field are both low drift tube (DT) chambers are used while in the endcaps cathode strip chambers (CSC) are more suitable due to the faster response and better radiation hardness. The spatial resolution of the DT are 100 μm×1 mrad while for the CSC it is 100-200 μm×10 mrad. Complementary to the DT and CSC systems, resistive plate chambers (RPC) are interleaved in the barrel and in the endcaps to provide a better timing resolution (1 ns) for triggering and providing a second independent measurement.

The DT chambers are composed of 3 superlayers (SL), each of them consisting of 4 sublayers of cells. The cells are the basic drift units and are tubes filled with Ar/Co_2 gas and with a charged wire inside acting as anode and cathode strips on the surface. The two outer SL have wires along the z direction while the inner SL is orthogonal to them

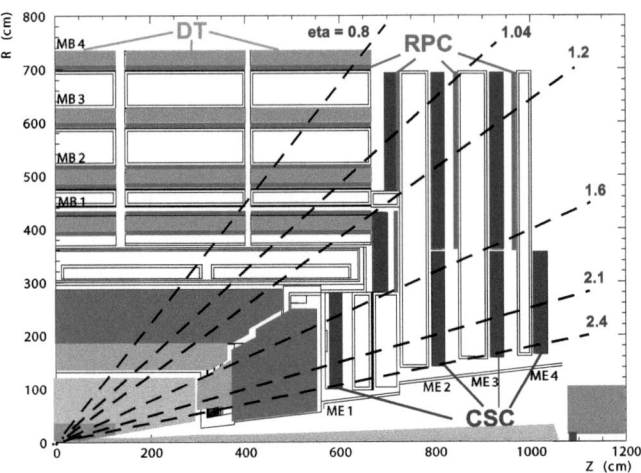

Figure 3.9: Schematic view of one quarter of the CMS muon system in the longitudinal plane with barrel and endcap detector layers composed of DT, RPC and CSC systems [74].

(fig. 3.10a) providing the $r - \phi$ and z coordinates, respectively. The honeycomb plate in the middle is attached to the iron yoke and supports the chamber serving as spacer. The outermost muon station only has two $r - \phi$ SLs and no z-coordinate measurement.

The CSC chambers are multi-wire proportional chambers consisting of 7 trapezoidal panels made of copper strips acting as cathode and 6 planes with anode wires in the gaps filled with gas (fig. 3.10b). The wires run in azimuthal direction and provide the radial coordinate and the cathode planes perpendicular to them the ϕ-coordinate.

The RPCs are avalanche detectors composed of two parallel high resistivity plastic plates, one acting as anode one as cathode, filled with gas. Metallic strips located on the exterior of one of the plates measure the electron avalanche produced in the gas by the passing muon. The estimated momentum from the pattern on the detecting strips is then used by the trigger.

3.2.7 Trigger

With almost a billion interactions per second, the roughly 100 million detector channels in CMS would produce around 100 TB of data per second. Both, the rate and the amount

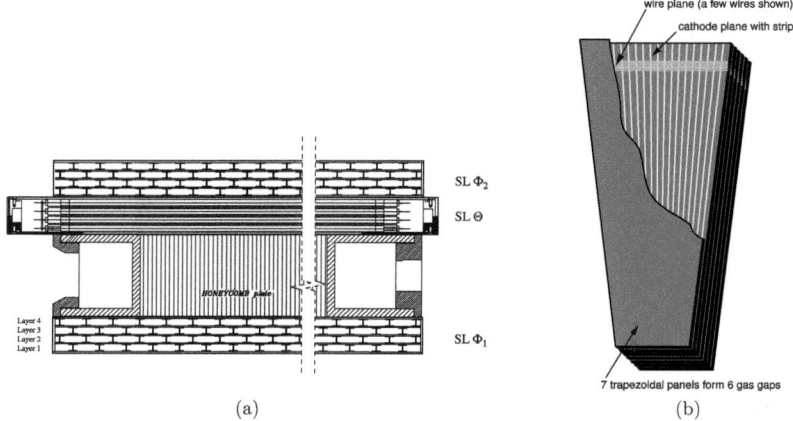

Figure 3.10: Layout of (a) a DT and (b) a CSC chamber in the muon system in the $r - \phi$ plane. SL Φ and SL Θ in the DT system refer to the 3 superlayers each composed of 4 sublayers. In the CSCs the 7 trapezoidal panels are visible along with the wires and strips in azimuthal and radial direction, respectively [74].

of data are too large to be handled and have to be reduced. The trigger system allows to decrease the event rate from the original 40 MHz down to around 400 Hz and a data rate of roughly 100 MB/s by preselecting interesting events and store the data in memory buffers so that several events can be processed together. The trigger systems consist of a Level-1 (L1) trigger which is based on programmable electronic and in a second step the High-Level-Trigger (HLT) based on a software system and a processor farm.

The L1 trigger uses a low level analysis with coarsely segmented data from the calorimeters and the muon system to decide wether an event is stored or not. The L1 trigger decision has to be received by the front-end electronics, where the high-resolution data is pipelined in 3.2 μs after the bunch crossing. Less than 1 μs of the total latency is allocated to the L1 trigger, the rest is needed for information transfer back and forth. The L1 decision is based on the presence of trigger primitive objects such as muons, electrons, photons and jets fulfilling certain energy or sometimes isolation criteria. These trigger primitives are generated with the local input of calorimeter deposits or hits in the muon chambers. Regional triggers combine the local information to create the objects. Then, global muon and calorimeter trigger information serves as input for the global trigger to decide if the event will be kept.

The full data of an accepted event by the L1 trigger comprises about 1 MB. It is read-out by the data acquisition system (DAQ) and passed to the online event filter system (processor farm) that executes the software for the HLT reducing the L1 output rate of 100 kHz to a few 100 Hz. The HLT makes use of the full event data, using the same software modules as in offline reconstruction. The trigger selections depend on the momentary luminosity scenario and can be adjusted at runtime. A HLT *trigger menu* contains a set of *trigger paths* where each path corresponds to the selection of one of the objects. Triggers can also be prescaled in order to reduce the output rate of a certain trigger path. A prescaled trigger will only be considered every n^{th} time, depending on the prescale factor n. Data accepted by the HLT are stored for offline data analysis. More information on the CMS HLT can be found in [75].

3.2.8 Computing

The amount of data which is stored for offline analysis is still very large, on the order of 5 PB per year at peak performance. To cope with this a worldwide distributed computing and storage system is used. The Worldwide LHC Computing Grid (WLCG) is able to distribute, process and store the data as well as support physics analysis and MC simulation production. The grid is segmented in hierarchical computing centers, the Tiers. A single Tier-0 (T0) center is located at CERN, seven Tier-1 (T1) centers at national computing facilities and on the order of 40 Tier-2 (T2) centers at institutes.

The T0 receives the data from the online system and copies it to permanent mass storage. This data format is called *RAW* and contains the full information delivered by the detectors and trigger systems. The RAW data is processed by algorithms from the modular CMS Software Framework (CMSSW), which is the software used in online and offline event processing, selection and analysis. The output from the prompt reconstruction of the RAW data are the *RECO* datasets and contain the high-level physics objects like muons or electrons plus the full record of the reconstruction hits and clusters used to produce them. The RAW and RECO data is then copied to the T1 centers and permanently stored. At the T1s the data is reconstructed for a second time with improved algorithms and/or calibrations and filtered datasets (Analysis Object Data, AOD) are produced containing only the high-level physics objects and sufficient additional information for kinematic refitting. The RECO and AOD datasets are transferred to the T2s where physics analysis can be performed. MC production can also be done at the T2s, which is then transfered to a T1 center for long term storage.

3.3 CMS Physics Measurements

A wide variety of physics studies is being carried out at CMS. Starting from the detailed studies of standard model physics with vector bosons and jets over to top- , b-physics and quarkonia, forward and small-x QCD physics up to the search for the eagerly awaited Higgs Boson and other physics beyond the standard model like supersymmetry (SUSY) and exotic phenomena, and last but not least the heavy-ion physics program. An overview as well as detailed information on these studies can be found in ref. [21] and the many publications of the CMS Collaboration [76].

3.3.1 Data Taking Periods 2010 and 2011 of CMS at LHC

The LHC started its operation in late 2009 with the first proton-proton collisions at an injection beam energy of 450 GeV. On 30 November a new world record was set by circulating beams at 1.18 TeV with first collisions in mid-December. First timing and alignment studies[3] could already be done with circulating beam previous to first collisions by means of beam-splash events where debris of particles from the impact of the beam on the collimators up- or downstream the detector leave traces in the detector. Shortly after the first collisions, π^0-resonances and the first dijet events were observed soon followed by the first J/Ψ candidate. After the winter shutdown, in March 2010, the first 7 TeV pp collisions took place. The calibrations and understanding of the detector were very advanced and the standard model particles could be 'rediscovered' at an enormous speed. Figure 3.11 shows this in an impressive way, where the many resonances in the dimuon mass spectrum can be seen for an integrated luminosity of $1.1\,\mathrm{pb}^{-1}$.

The total integrated luminosity over the 2010 data taking year at $\sqrt{s} = 7\,\mathrm{TeV}$ recorded by CMS was $43\,\mathrm{pb}^{-1}$ out of $47\,\mathrm{pb}^{-1}$ delivered by LHC, which is roughly 92% data taking efficiency. Only certified data is used for analysis, where each data block only enters the certification list if all sub-detectors and trigger systems are marked as in good conditions. In total $(36.1 \pm 1.4)\,\mathrm{pb}^{-1}$ [77, 78] were certified for analysis, where the peak luminosity reached at the end of the period was $\mathcal{L} = 10^{32}\,\mathrm{cm}^{-2}\mathrm{s}^{-1}$.

After the winter shutdown, the LHC resumed its operation at 7 TeV delivering to CMS a total luminosity of $6.095\,\mathrm{fb}^{-1}$ out of which CMS recorded $5.561\,\mathrm{fb}^{-1}$ 2011 data with stable beams [79]. More information on the LHC performance parameters during the 2010 and

---
[3]Previous to LHC operation many calibration and commissioning studies had been done with cosmic ray muons.

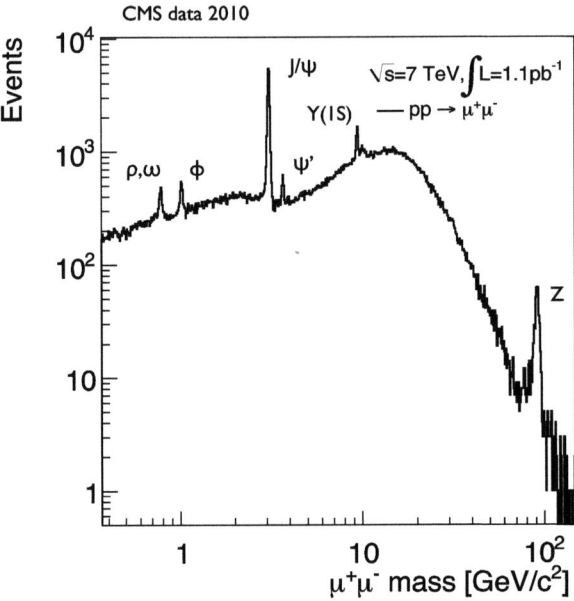

Figure 3.11: Resonances in the dimuon mass spectrum measured with 2010 CMS data at an integrated luminosity of $1.1\,\text{pb}^{-1}$.

2011 runs can be found in ref. [80] and [81]. In 2012 it was decided to increase the energy to 8 TeV [81] and as of June 2012, $6.65\,\text{fb}^{-1}$ have been delivered to CMS and $6.15\,\text{fb}^{-1}$ recorded, outrunning the total 2011 luminosity in only a few months.

In this thesis, two different data samples are used. For the early analysis concerning the energy flow, multiplicity and diffractive measurements (chapters 6 to 8) the 2010 data sample was used while the W and Z signal selection analysis (chapter 5) and the measurement of the transverse momentum spectrum (chapter 9) were redone with a subset of the 2011 data sample corresponding to an integrated luminosity of $1.5\,\text{fb}^{-1}$.

3.3.2 CMS Measurements with W and Z Bosons

W and Z bosons with leptonic decays are one of the best understood processes at hadron colliders. Due to this and the copious production they can be used for performance studies, understanding efficiencies, resolution, energy scale, triggers and general calibration and understanding of the detector response.

Moreover, with the current very accurate knowledge of the theoretical cross section, the proton or even more precisely the parton luminosity can be determined [82].

For many new physics searches, W and Z bosons form part of the major background, e.g. for Higgs boson, single top, SUSY searches, and others, and thus, can be used for background estimation and subtraction methods.

Precision measurements of the standard model parameters, which can only be obtained by experiment and are input to the SM, serve to either validate and improve the current models or to detect inconsistencies which indicate the presence of new physics. For example the Higgs mass, the only parameter which has not been measured, can be constrained by the top and the W mass related via radiative corrections. The W mass is a limiting factor there. For a precise M_W measurement it is necessary to have a good knowledge of the production and decay kinematics. Thus, it is important to study the transverse momentum and rapidity spectra of the W, which provide a good understanding of the W process. Particularly, it provides constraints on quark and gluon PDFs[4]. PDFs are used for cross section predictions for any other process with the same partons (i.e. quarks and gluons) involved in the production.

In the following, a short summary of some of the measurements related to W and Z physics which have been performed with CMS data is given.

Inclusive and differential cross sections

The first measurements performed at CMS were the inclusive cross section measurements [83] which were later updated with the full 2010 data sample[84]. W^+ and W^- as well as the ratios W^+/W^- and W/Z are measured, and through W/Z the width of the W is indirectly determined, using previous Z width measurements [85]. The results are in agreement with SM NNLO cross section calculations based on recent PDFs and with the ATLAS measurements [86].

Moreover, differential cross sections were measured which provide tests for perturbative QCD, constraints on PDFs, and, in the Drell-Yan [87] differential cross section measurement $d\sigma(Z/\gamma*)/dM_{ll}$ [88], indications for possible dilepton resonances at high mass. The

---

[4]PDFs do not limit the W mass measurement in principle, but due to the restricted acceptance of the detector a sensitivity to longitudinal boosts is inserted

Z differential cross sections $d\sigma/dp_T$ and $d\sigma/d|y|$ [89] were done with the full 2010 data set. In the rapidity measurement ($|y| < 3.5$) an overall agreement was found with the models taken into account, while for the p_T distributions no model agreed in the entire spectrum range ($p_T < 600\,\text{GeV}$).

W charge asymmetry

Due to the composition of the proton (uud) with more $u-$ than $d-$ valence quarks and the dominant W productions $u\bar{d} \to W^+$ and $d\bar{u} \to W^-$, more W^+ than W^- bosons are produced. The ratio measured recently with CMS is 1.43 ± 0.05 and in agreement with SM predictions [84]. The excess of W^+ over W^- is also rapidity dependent according to eq. (2.26). For small rapidities the parton momentum fractions x_1 and x_2 are small and similar, corresponding to mostly sea $q\bar{q}$ annihilation. At large rapidities the x values become very different favoring the production with a large x valence quark and a small x sea quark involved. Thus, at forward rapidities the W^+ production with the larger x u quark compared to the d quark is favored, leading to an asymmetry in the rapidity distributions. Due to the missing neutrino information, experimentally the asymmetry is observed via the lepton pseudorapidity asymmetry which is defined as

$$A(\eta) = \frac{dN/d\eta(l^+) - dN/d\eta(l^-)}{dN/d\eta(l^+) + dN/d\eta(l^-)}. \tag{3.1}$$

The parity violation in the weak interaction enhances the asymmetry between the leptons. The emission of left-handed leptons in the same direction relative to the W rest frame as the left-handed quark which created the W is favored, and correspondingly for the right-handed anti-fermions, leading to an asymmetry in the lepton distribution. The asymmetry was measured with the 2010 CMS data set for muon and electron decays and can be used as input for PDF global fits [90, 91, 92].

W polarization

W bosons with large transverse momentum are predominantly polarized left-handed due to the parity violation in the production process in combination with the dominance of the valence quarks in the protons [93, 94]. The spin conservation and the larger x value for the quarks compared to the antiquarks lead to the left-handed polarization for both the W^+ and W^- bosons. The leptons l^- and ν (l^+ and $\bar{\nu}$) are predominantly emitted forward relative to the W boson direction, which results in an increased (decreased) p_T.

Thus, a handle to discriminate other searches with a predominant W + jet background is given. The polarization was measured with CMS and the SM predictions were found to be confirmed [95].

W and Z bosons in association with multiparton interactions

Measurements with soft QCD final states have been carried out to study the non-perturbative effects of the LHC physics. The understanding of these processes is still in its very beginning, the models describing them rely on phenomenological descriptions and consequently on the tuning of data. Thus, the measurements are important for a better understanding of the non-perturbative physics which is an integral part of SM precision measurements as well as searches for new physics. This can range from the determination of the losses involved with lepton isolations, to reconstruction efficiencies in e.g. $H \rightarrow \gamma\gamma$ events where the vertex is determined from the UE [96]. The involved studies used for the tuning have mostly be done with minimum-bias data and events with high p_T jets. CMS studies of the underlying event have been done at central rapidities [96, 57] and forward rapidities [97] as well as studies with charged particle multiplicites [98], rapidity and transverse momentum distributions of charged hadrons [99, 100, 101, 102] and many others (for a full record of CMS measurements concerning underlying event and soft QCD studies see [103]).

The underlying event activity is expected to depend on the scale of the interaction [104, 105] which makes it interesting to study the UE structures in association with W or Z bosons. This is especially true, because the final states in W and Z events with subsequent leptonic decay lack of QCD final state radiation and for muons the EWK radiation is negligible. Thus, the hard interaction can be well separated from the UE and comparisons to different processes, like those including FSR can be done. The underlying event in the Drell-Yan process was studied in the muonic final state [106]. The influence of multiparton interactions on the charged particle multiplicities and the forward energy flow as well as correlations between them is studied in W and Z events and will be presented in this thesis. Finally, with a better understanding of the underlying soft physics, diffractive processes can be studied, eventually leading to a better understanding of this still not well understood phenomenon.

Chapter 4

Muon Identification

Many of the potential new particles which are being searched for with CMS contain leptons in their final states. Leptons have a very convenient signature which allows to distinguish the new particle from the large background and to trigger on. In CMS, the cleanest and easiest to detect lepton with energy in the GeV range is the muon. Due to its characteristic of crossing large detector regions with a minimal interaction, it is easily discriminated from the other decay particles. CMS was designed to exploit this fact, and it is thus natural to choose the muon decay channel to select W or Z bosons.

The basic ingredient for all analyses in this thesis are high energetic muons. In this chapter, the details for the muon identification in CMS are explained with an emphasis on the three most important performance criteria: a high muon selection efficiency, compromising with a low fake-rate, and a good momentum resolution.

For the startup phase of data taking, the major focus is put on the understanding of the detector with clear separation of sources for systematic errors. For this reason, in the beginning the acceptance for the muons is restricted to the barrel. The final studies were presented in combination with another analysis done with the electron channel and therefore the acceptance restriction was chosen to be the ECAL barrel, which is not identical to the muon system barrel.

For the analysis in the later stage of data taking, in the 2011 runs, the acceptance is expanded to the endcap regions. In the following sections the standard muon reconstruction algorithms used at CMS are presented together with the basic kinematic variables. Afterwards a short overview of the muon resolution performance of CMS is given and the selection variables and triggers are introduced and discussed, establishing the basic muon selection cuts used in the following analyses.

4.1 Muon Reconstruction Algorithms

The standard muon reconstruction algorithms used at CMS take advantage of the different detector components, which allow a robust and accurate measurement of the muon properties. By combining the silicon tracker and the muon system information, an excellent performance can be obtained. Three different muon reconstruction objects are defined: *standalone muons* which use only muon system information, *global muons* with a standalone muon track from the muon chambers matched to a tracker track from the inner tracking system and *tracker muons* with a tracker track which has at least one compatible segment in the muon chambers. The following description summarizes the three reconstruction methods explained in detail in [74], where more information about the algorithms and their performance can be found. The standard muon objects used in CMS contain a collection of all three algorithms (without double-counting) using as a default kinematic variable the best reconstruction available.

The standalone muon algorithm starts with the reconstruction of hit positions in the DT, CSC and RPC. Hits within each DT and CSC are matched to segments which are on turn then matched to generate seeds. The seeds are the starting point to locate compatible hits in subsequent detector layers, working from inside out. The track candidates are then extrapolated to the point of closest approach to the beam line and a vertex constraint is applied.

For the global muon candidate, the starting point is a standalone reconstructed track. The trajectory is then extrapolated from the innermost muon station to the outer tracker surface, taking into account the muon energy loss in the material and the effect of multiple scattering. A region of interest is defined in which a subset of tracker tracks is selected and more stringent spatial and momentum matching criteria are applied. After a global refit one global muon track remains identified as global muon.

In some cases the quality of the muon tracks in the muon system is not good enough for a starting point. This can happen when muons are lost due to geometrical effects like gaps between the wheels. The tracker muon reconstruction algorithm starts with all the silicon tracker tracks matching them with at least one segment in the muon system. No combined silicon-muon track fit is performed. Due to the very loose association between segments and tracker tracks this complementary algorithm is not suitable without any further quality requirements.

Figure 4.1 shows the pseudorapidity and transverse momentum spectrum for muons reconstructed with the three different algorithms compared to the generated distributions

Muon Identification

for a W $\to \mu\nu$ MC sample. The line representing the reconstructed muons comprises the entire collection. No cuts have been applied to the muons after the reconstruction. The "ears" in the endcap region of the pseudorapidity distribution clearly show that the tracker muon algorithm should not be applied without further quality cuts. The same conclusion is drawn from the p_T spectrum where we can see that these "fake" muons are mostly muons with low p_T. The standalone and the global muon algorithm have a much smaller fake-muon component. The fake-muons reconstructed with the tracker algorithm are also visible in the ϕ-distribution, but no distinguished direction is observed.

Figure 4.2 shows the MC efficiency to reconstruct a generated muon as a function of the pseudorapidity with each one of the algorithms and for all of them combined. The efficiency here is defined as the number of reconstructed over generated muons with a preselection cut of $p_T > 20\,\text{GeV}$ and $|\eta| < 2$, which was applied to reduce the large fake rate of muons at low p_T and to constrain the geometry to the pseudorapidity region of the muon system. All the algorithms obtain an efficiency of more than 90% as a function of the pseudorapidity. The geometry structure of the tracking and the muon system with their gaps become visible in the different plots. The small drop in efficiency for the tracker muons around 0 is due to the tracking system geometry. The tracker is made of two half-barrels joined at $|\eta| = 0$. The standalone muons have a dip around $|\eta| \sim 0.3$, 0.8 and 1.2. This can be ascribed to the discontinuity between the wheels making up the muon system. The global muons have the overlap of drop in efficiency from the tracker and standalone muons.

The geometry also becomes visible in the efficiency as a function of ϕ shown in fig. 4.3. There the periodic structure reflects the gaps between two neighboring sectors. All of the algorithms have an efficiency larger than 97% in any ϕ-region.

4.2 Muon Momentum Resolution

The momentum resolution can be described with the Glueckstern equation [107]

$$\frac{\delta p_T}{p_T} = \frac{0.0136}{\beta BL}\sqrt{\frac{x}{X_0}}\sqrt{\frac{4A_N}{N}} \oplus \frac{\sigma p_T}{0.3BL^2}\sqrt{4A_N}$$

where B is the magnetic field, L the length of the tracking system, $\frac{x}{X_0}$ the thickness of the scattering medium in radiation lengths, N the number of hits, A_N a function of N and σ the individual error. The first term corresponds to the contribution of multiple

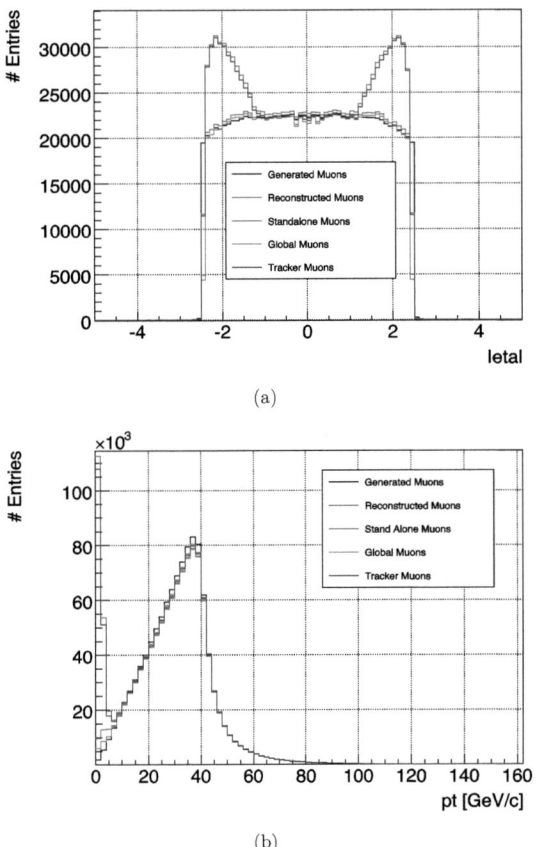

Figure 4.1: (a) Pseudorapidity and (b) transverse momentum distribution for muons reconstructed with the standalone, the global and the tracker muon reconstruction algorithm, compared to the generated distributions for a $W \to \mu\nu$ MC sample. The line for the reconstructed muons corresponds to the entire muon collection comprising all three algorithms.

scattering and is constant with respect to p_T. The second term is directly related to the measurement precision. In the muon system the resolution is dominated by multiple scattering in iron. It is almost constant up to roughly 100 GeV. Above, the second term becomes important. The resolution in the tracking system is dominated by the second term. With increasing p_T this term becomes larger but can be balanced by a longer path

Muon Identification

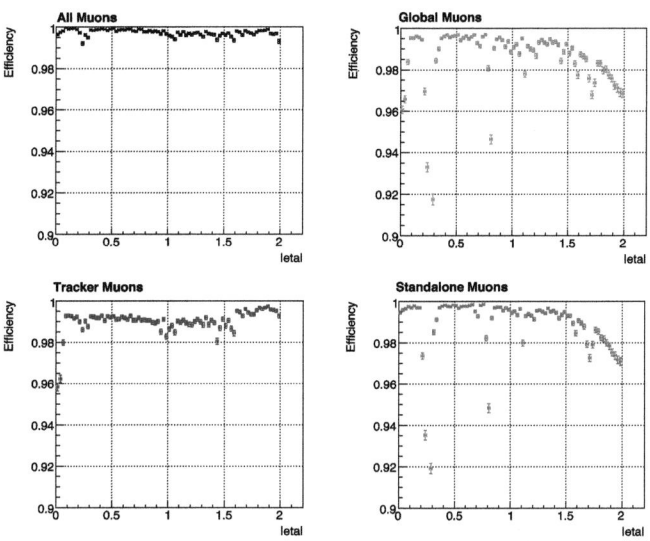

Figure 4.2: Efficiency as a function of η for the three different muon reconstruction algorithms and for the combined muon collection. A preselection cut of $p_T > 20\,\text{GeV}$ and $|\eta| < 2$ is applied.

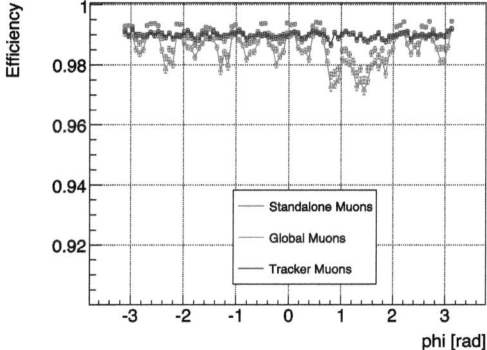

Figure 4.3: Efficiency as a function of ϕ for the three different muon reconstruction algorithms. A preselection cut of $p_T > 20\,\text{GeV}$ and $|\eta| < 2$ is applied.

length L. Therefore, the best resolution can be obtained by combining tracking and muon system. Nevertheless, this becomes relevant only above a few hundred GeV. Below, where the muon system resolution is large and does not contribute to the combined one, the tracking and the combined system deliver practically the same resolution. This can be seen in fig. 4.4 where the resolution is shown as a function of p for the three systems and for the barrel and endcaps as expected from MC simulation. The resolution in the barrel is better than in the endcaps due to a larger amount of material (additional 60 radiation lengths) in the latter. The degradation of resolution with increasing pseudorapidity can be seen in fig. 4.5 for global and standalone muons.

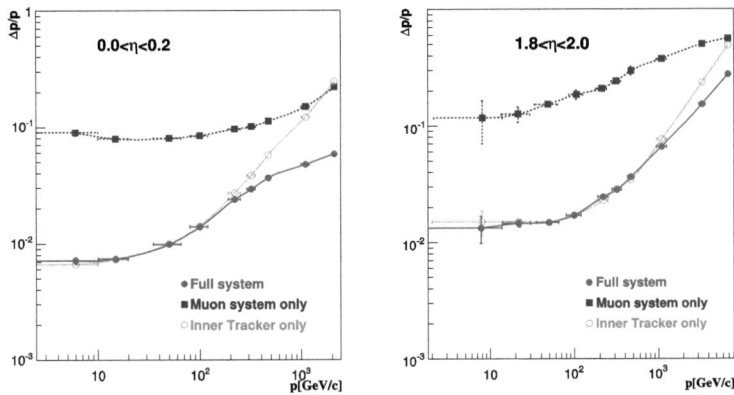

Figure 4.4: Muon momentum resolution as a function of p for the three standard CMS muon reconstruction algorithms for barrel (left) and endcap (right) in the CMS detector [74].

During the first periods of data taking at 7 TeV, the transverse momentum resolution was measured using low-mass resonances and found to be in very good agreement with the MC predictions, the largest difference being 5% in the transition region between barrel and endcaps. It was found to be roughly 1% in the barrel and 2% in the endcaps for p_T < 10 GeV [108].

4.3 Muon Selection

To establish a set of cuts which optimize the selection of muons coming from a W or a Z boson, two important criteria have to be taken into account: a reconstruction efficiency as high as possible and a fake-rate as low as possible. To study the fake-rate for muons in W events, the reconstructed muons from MC simulations are classified into *fake* and *good* muons according to their provenance. Each reconstructed muon is matched to

Muon Identification

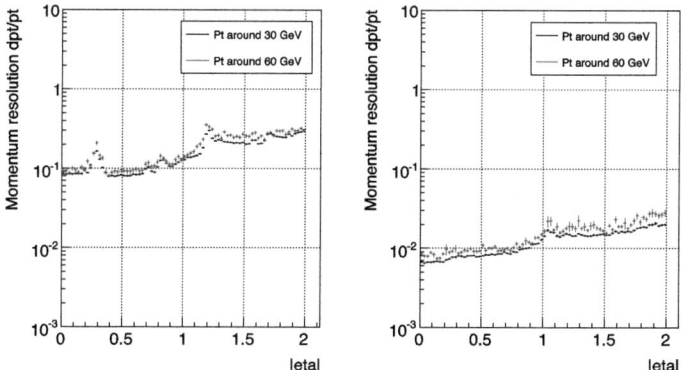

Figure 4.5: Muon momentum resolution as a function of η for standalone (left) and global (right) muons in the CMS detector.

its corresponding generated particle on an event-by-event basis and classified as good if it matches within a ΔR cone [1] of 0.05 with a simulated muon coming from a W and as fake otherwise. This means that real muons can be included in the fake category. Fake muons contain *punch-through hadrons*, *decay-in-flight muons* and *duplicates*. Punch-through hadrons are mostly pions and kaons traversing the detector and leaving a trace in the muon chambers. Decay-in-flight muons are true muons originating from either light flavor or heavy flavor hadrons. Muons from the light flavor category include decay-muons from pions and kaons but also, to a less extent, from a nuclear interaction in the detector or a calorimeter shower. Heavy flavor muons result from any other hadron not included in that category. Finally, duplicates result from miss-measurements leading to more than one muon reconstructed out of one.

To establish a set of cuts based on a simple cut-and-count method a MC sample based on the standard PYTHIA simulation for W bosons decaying to muons with ideal detector conditions is taken. In this sample, roughly 4% of all reconstructed muons correspond to the fake category, mainly consisting of decay-in-flight muons. Additionally to the basic preselection cuts of $p_T > 20\,\text{GeV}$ and $|\eta| < 2$ mentioned before, the global muon flag is required to ensure good quality muons and the acceptance is restricted to the barrel ($|\eta| < 1.48$). The p_T cut leaves 82% of the good muons, while only 2% are fake. The acceptance restriction to the barrel cuts 50% of the good events and leaves 1% fakes. This

---

[1] $\Delta R = \sqrt{\Delta\phi^2 + \Delta\eta^2}$ where $\Delta\phi$ and $\Delta\eta$ are the distance between the cone axis and the energy deposit position in ϕ and η angle, respectively.

loss is acceptable due to the large rate of W and Z production at the LHC already in the first data taking period. The restriction to the barrel on the other hand simplifies the understanding of the detector and the different quality criteria variables at the startup phase.

4.3.1 Muon Isolation and Quality Selection Cuts

One of the most useful characteristics of some heavy objects like W and Z bosons is that their muons are only surrounded by particles from uncorrelated processes from the underlying event or pileup, which leave no or very small traces in the detector. The muons are called to be *isolated*. This characteristic is a very powerful tool to discriminate between the muons from a W or Z and those from hadronic events, called QCD events. Muons in QCD events are typically surrounded by particles leaving traces in the tracking system and energy in the calorimeter. The standard isolation algorithm used in CMS counts the total energy above some threshold deposited in a cone around the muon track and subtracts the corresponding muon energy (veto-value) (cf. fig. 4.6).

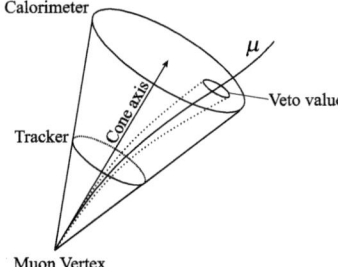

Figure 4.6: Sketch of the isolation cone around the muon trace illustrating the muon isolation algorithm.

The energy of the particle activity inside the cone can be calculated using the tracking system or the calorimeters, or combining them. The isolation cut efficiency is shown in fig. 4.7 for several versions for the energy counting and for a cone of $\Delta R = 5$ radians for a simulated W $\rightarrow \mu\nu$ sample. The isolation in fig. 4.7a is defined as the sum of all the energy deposits in the ECAL inside the cone and removing the energy within a veto-value cone of 0.07 radians. The corresponding distribution for deposits in the HCAL is shown in fig. 4.7b where the veto-value cone is 0.1 radians. In fig. 4.7c the isolation is defined as the sum of the transverse momentum of all the tracks reconstructed inside the cone with $p_T > 1\,\text{GeV}$. The veto-value cone is of 0.01 radians. The sum is then divided by the muons transverse momentum. This relative quantity delivers a more homogeneous variable than the absolute value. Finally, fig. 4.7d shows the sum of all of them, i.e the sum of the tracks momenta plus the energy deposits in ECAL and HCAL, inside the cone.

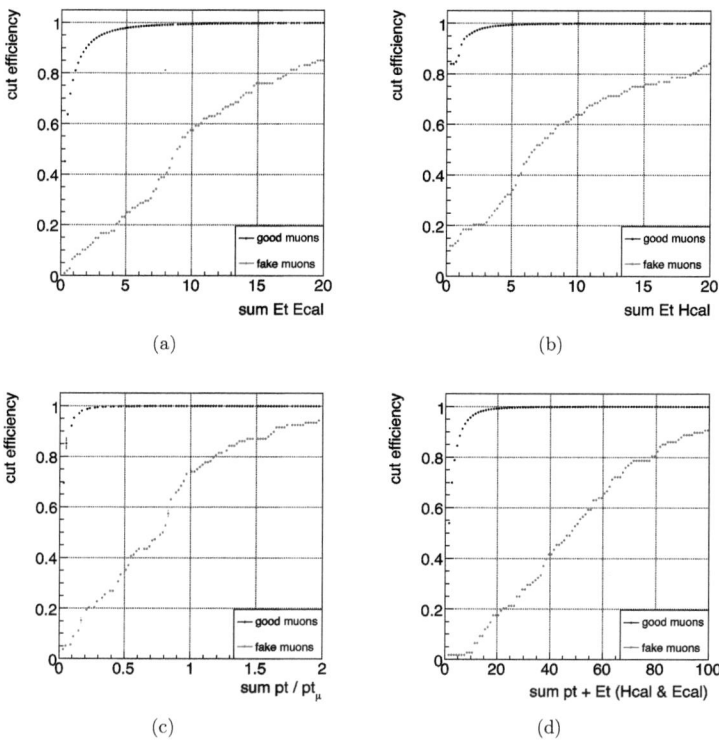

Figure 4.7: Cut efficiency for four different isolation variables for an isolation cone of $\Delta R = 5$ radians for fake and good muons as expected from MC simulation for a sample of global muons with a preselection cut of $p_T > 20\,\text{GeV}$ and the restriction of the acceptance to the barrel ($|\eta| < 1.48$).

All of the isolation variables present a very good rejection performance, with a slightly better performance for the combined isolation variable. The rejection power for cone sizes 0.3 and 0.5 are very similar. In the startup phase a cone size of 0.5 was used, switching then to the smaller size of 0.3 to account for the increasing pileup activity.

As seen before the requirement of the global muon tag ensures the best quality of the momentum reconstruction of all three muon reconstruction algorithms. Additionally to this several track quality criteria can be applied:

- a good χ^2 value for the combined global fit
- the number of hits associated to the tracker, the muon and the global track
- the number of layers in the muon tracking system crossed by the muon
- the impact parameter d0 of the track in the transverse plane with respect to the primary vertex

The χ^2 value and the number of hits and layers ensure a good momentum reconstruction and further reduction of the fake-rate, especially of punch-through particles, while a small impact parameter rejects a residual contamination of cosmic muons. Figure 4.8 shows the number of hits in the tracker, the normalized χ^2 value, the impact parameter and the isolation for good and fake muons as obtained from MC simulation and with the cuts applied in successive order. The impact parameter is corrected for the beam-line offset with respect to the beam spot[2].

In table 4.1 the fraction of surviving good and fake muons with $p_T > 20\,\text{GeV}$ and $|\eta| < 1.48$ is shown in successive steps for the most important selection cuts mentioned above.

Table 4.1: Fraction of surviving good and fake muons with $p_T > 20\,\text{GeV}$ and $|\eta| < 1.48$ in successive steps for various muon selection cuts obtained from a $W \to \mu\nu$ MC sample.

	good muons	fake muons
global μ	99.25%	1.50%
hits in tracker > 10	99.06%	1.39%
$\chi^2 < 10$	97.71%	0.64%
Iso < 10	94.34%	0.02%

Summarizing, the following muon selection cuts are used for the analyses in this thesis:

- global muon
- transverse momentum $p_T > 20$ GeV
- pseudorapidity $|\eta| < 1.48$ (barrel acceptance)
- number of hits in tracking system ≥ 10
- global fit normalized $\chi^2 < 10$
- impact parameter $|d0| < 0.5$

---

[2]The beam spot is the luminous region produced by the collisions of proton beams

Muon Identification

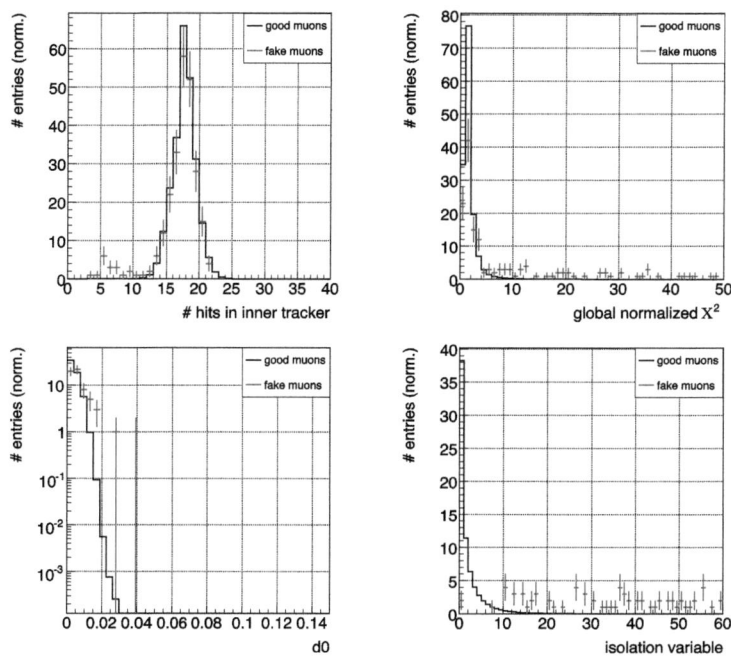

Figure 4.8: Muon selection variables for good and fake muons as obtained from MC simulation for a sample of global muons with a preselection cut of $p_T > 20\,\text{GeV}$ and the restriction of the acceptance to the barrel ($|\eta| < 1.48$). The events are normalized and only shapes are being compared.

- combined isolation $< 10\,\text{GeV}$ (cone 0.5)

These cuts were used in the beginning of the first data taking period with CMS. With increasing luminosity and collected data some of the cuts were tightened and a few were added, increasing the purity of the selected sample. The acceptance of the muons was extended to include the endcaps. For the 2011 data taking period the following cuts were used:

- global muon and tracker muon tag
- transverse momentum $p_T > 20\,\text{GeV}$
- pseudorapidity $|\eta| < 2.1$
- number of hits in tracking system ≥ 10

- number of hits in muon system ≥ 2
- number of hits in pixel tracker system ≥ 1
- global fit normalized χ^2 <10
- impact parameter $|d0| < 0.2$
- relative combined isolation < 0.15 (cone 0.3)

4.3.2 Muon HLT Triggers

As described in section 3.2.7 a trigger system has to be used to select only the interesting events out of the huge amount of data. During the first two years of data taking at the LHC the running conditions were changing continuously and accordingly the trigger system too. A trigger menu was defined providing different triggers suitable for the different analyses. With increasing luminosity the conditions for the muon triggers had to be adapted and the transverse momenta thresholds were risen. The muon triggers used in this thesis depend on the data taking period. For 2010 data a trigger with final p_T threshold of 9 GeV was used, while for 2011 data the threshold was increased to 17 GeV.

Chapter 5

W and Z Boson Event Selection

The precise measurement of distributions in W and Z events, such as the forward energy flow (chapter 7), or the W (Z) transverse momentum (chapter 9), relies primarily on an experimentally clean and well understood selection with a good signal to background ratio. Especially the Z boson, but likewise the W, has a very simple signature and is produced abundantly at the LHC. This allows for an excellent signal to background ratio and very small impact of insufficiently understood processes, such as QCD related ones, with only few selection cuts. Nevertheless, the selection variables need to be under control, and the detector performance and object reconstruction, such as muons, electrons and others, has to be understood to exclude biases in the measured distributions. The main constituents for the W and Z selections are, besides high energetic and isolated muons, the reconstruction of the neutrino in the W events, jets for the p_T studies and the vertex reconstruction for the studies in chapters 6 to 8. In the following sections these components will be described and the results obtained with CMS data are presented.

5.1 Missing Transverse Energy

The leptonic W decay channel implies neutrinos in the final state. Neutrinos cross the CMS detector without leaving any trace. Nevertheless, some of the missing information can be regained by adding vectorially the energy of all final state particles in the transverse plane. Due to the hermetic design of the CMS detector this should sum up to zero when no particle is missing and equal the negative neutrino transverse energy in a W event. Unfortunately, the missing transverse energy (MET) is very sensitive to detector imperfections and mis-calibrations. Another source of missing E_T mis-measurements are beam related background and cosmic muons. All these influences can broaden the resolution of the real neutrino transverse energy or even result in a large fake MET. In spite of this the MET variable is one of the key variables for the W reconstruction and a very powerful discriminating tool against QCD and Drell-Yan background. Several algorithms exist to obtain the sum of the transverse energy. The most classical approach is to simply sum the energies deposited in the calorimeter towers. The calorimetric Met (*caloMet*) algorithm [109] sums all energies above noise threshold in the pseudorapidity region $|\eta| < 5$. A correction from the measured momentum is applied for muons as they leave only a minimal part of their energy (around 2 GeV) in the calorimeter. The energy corresponding to the measured momentum is added and the minimum energy deposit in the calorimeter is subtracted.

Additionally to simply summing the energies, all charged hadrons can be corrected by their measured momenta, taking advantage of the good resolution of the CMS tracking system. The track corrected algorithm (*tcMet*) [110] corrects the caloMet energy by the measured momenta in the inner tracker, subtracting the corresponding energy in the calorimeter. For muons the correction is very similar as the one applied in the caloMet algorithm. The major improvement of the tcMet is a better control of the tail of the MET distribution.

An extension to the tcMet is the reconstructed MET based on the particle-flow algorithm (*pfMet*) [111]. The particle-flow technique is based on the unique identification of all stable particles involved in the event, making use of the combined information of all the subdetectors in CMS. The list of particles contains muons, electrons, photons, and neutral and charged hadrons. No additional correction is applied to the muons, as the algorithm itself accounts for the several detector components.

In the startup phase of data taking the mentioned influences on the MET measurements like beam related background and noisy cells or likewise first have to be studied and implemented in the simulations. For this reason, beforehand, the option of using only the hard objects in the event to reconstruct the missing E_T was studied in MC simulations. In the case of the W $\rightarrow \mu\nu$ events this includes muons and jets with minimal transverse momenta of $p_T > 20\,\text{GeV}$ and $30\,\text{GeV}$, respectively. The comparison of the reconstructed MET with this method and with the standard caloMet algorithm is shown in fig. 5.1 for W $\rightarrow \mu\nu$ events with zero and one jets.

For the events with no additional jet, both algorithm show a similar agreement with the generated distribution. In the events with one reconstructed jet, the shapes are much broader and the discrepancy is much larger. This can be explained by the additional jet

W and Z Boson Event Selection

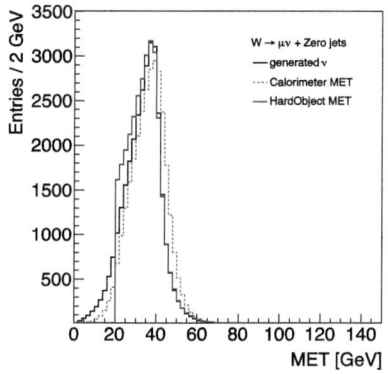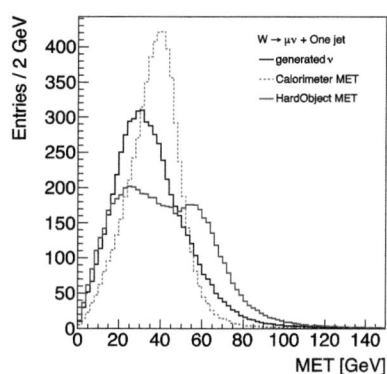

Figure 5.1: Comparison of the generated neutrino transverse momentum with two different MET reconstruction algorithms for W → $\mu\nu$ events with zero jets (left) and one jet (right).

in the event, which is reconstructed much less precise than the muon, and hence, will broaden the spectrum. Additionally, the shape of hard object MET has a distortion in the shape around 50 GeV, which comes from the cut-off in p_T applied to select the hard objects, and which are then added vectorially. The peak of the caloMet on the other hand, is shifted towards higher energy. This can be explained by the threshold applied at the tower level to select energy deposits.

Thus, we can conclude that the caloMet delivers the better performance and the hard-object MET was only kept as a backup solution for a "worst-case-scenario" in the startup phase. However, with the long preparation period of CMS during the repair times of LHC many things were already understood, calibrated and adjusted when the proton-proton collisions started and no such backup solution was needed. The particle-flow algorithm soon established as the standard and CMS-wide used algorithm with a very stable performance. Several performance studies were done with $\sqrt{s} = 7$ TeV data, with minimum-bias and jet events [112], and with events containing electroweak bosons [113]. It was found that the agreement between data and MC is good and that the inclusion of the tracking system to estimate the energy improves the performance significantly. Figure 5.2 shows the missing E_T distributions for all three algorithms for W → $\mu\nu$ candidate events in 2010 pp collision data and for an integrated luminosity of 246 nb^{-1}. The broader resolution of the caloMet compared to the tcMet and pfMet is clearly visible, while the latter two are comparable. For the W → $\mu\nu$ candidate event selection in 2010 and 2011 data, particle-flow MET was used in all following analyses.

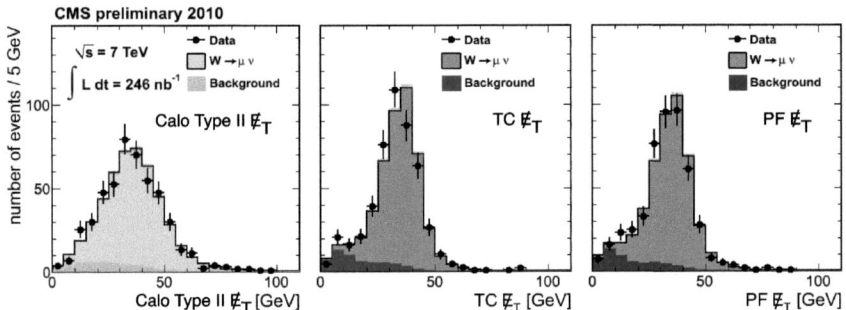

Figure 5.2: Missing E_T distributions for caloMet, tcMet and pfMet, from left to right for $W \to \mu\nu$ candidate events in spring 2010 CMS data for a collected luminosity of 246 nb^{-1} [113].

5.2 Jets

Jets will be an important component in the analysis to determine the transverse momentum spectrum of the W boson (chapter 9). Here, a short overview of the different reconstruction methods and the performance will be given.

Analogous to the MET reconstruction, jet reconstruction at CMS is based on three different types: calorimeter-only (*caloJet*), track-corrected (*Jet-Plus-Track* or *JPT*), and particle-flow based (*pflow*) jets. After the object (particles or calorimeter towers) identification as explained in the previous section, the anti-kt clustering algorithm with cone size $\Delta R = 0.5$ is applied [114]. Performance studies have been done with 2010 data for all three types, and like for the MET studies it was found that the types including tracking information have a significantly better performance. Overall, a very good description of the jet properties was achieved by the CMS simulation [115]. The two most important criteria for a good jet performance are the jet energy response and the resolution.

The jet energy resolution has been studied with a data sample of 35 pb^{-1} integrated luminosity for all three jet types and close agreement was found between data and MC [116]. In chapter 9, a more detailed discussion about the resolution, in the context of the Z and W p_T spectrum, will be given.

The jet energy is affected by the non-linear and non-uniform response of the CMS calorimeter. Additionally, electronic noise and pileup can increase the real jet energy. Thus, the reconstructed jet energy will differ from the generator level energy. This is corrected by multiplying the "raw" reconstructed jet energy by a factor determined from MC simulations. This approach is validated by the good agreement found between data and MC. On top of the MC truth energy calibration a residual correction determined from data is applied. From measurements done at the beginning of the data taking an uncertainty of 3 – 6% of the overall jet energy scale for a jet p_T from 15 GeV to 2 TeV for pflow jets was found [117]. In fig. 5.3 the total jet energy scale uncertainty is shown as a function of the jet p_T for two different pseudorapidity values and for the three jet types. For both η values the uncertainty raises for low p_T and for high p_T.

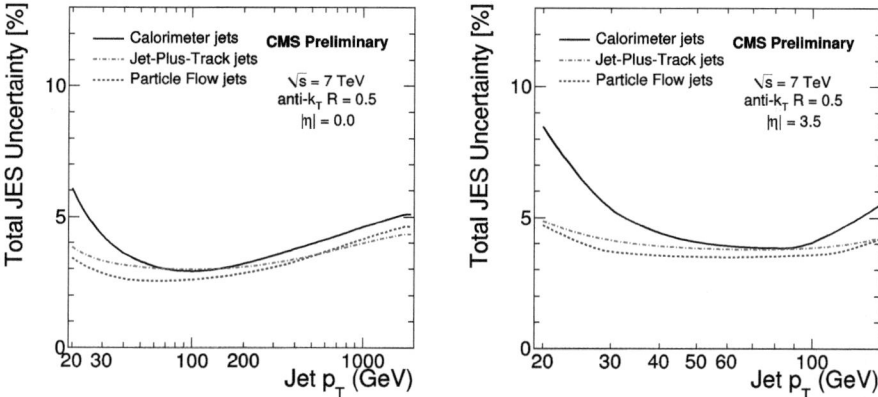

Figure 5.3: Total jet energy scale uncertainty as a function of the jet p_T for $\eta = 0$ (left) and $\eta = 3.5$ (right) for the three jet types used at CMS and for $\sqrt{s} = 7$ TeV data [117].

In this thesis, pflow jets are used with the anti-kt clustering algorithm and with a cone size of $\Delta R = 0.5$. The jet energy corrections are applied as explained above, including the residual calibration on top of the MC correction for data. A transverse momentum threshold of 30 GeV is chosen to avoid the larger systematic errors below that value.

5.3 Vertex Reconstruction

The reconstruction of the primary interaction vertex (primary vertex) of an event is not only crucial for most of the analyses at CMS but particularly important for the diffractive physics analysis described in chapter 6. As will be discussed in more detail later, the identification of events with more than one primary vertex is essential to filter out the events with pileup and preserve the gap signature of diffractive events.

The starting point for the vertex reconstruction is a collection of reconstructed tracks which fulfill quality criteria like impact parameter significance with respect to the beam line, number of hits in the pixel and the strip detector and a good normalized χ^2. The selected tracks are then grouped if their z-coordinates closest to the beam line are separated by less than 1 cm from their nearest neighbor. A fit is applied where each track obtains a weight between 0 and 1 corresponding to its compatibility with the vertex based on several parameters like position, covariance matrix and fit parameters, including number of degrees of freedom of the vertex (N_{dof}). The number of degrees of freedom is defined as $N_{dof} = 2 \cdot \Sigma_{i=1}^{N_{tracks}} w_i - 3$, with w_i the weight of the ith track. The weight w_i is close to 1 if the track is compatible with the common vertex and close to zero for the opposite case. The N_{dof} therefore corresponds, roughly speaking, to the number of involved tracks and will be used later to study the identification efficiency and rejection power of pileup events in diffractive W and Z bosons. More details about the vertex reconstruction algorithm are given in [118].

The performance of the vertex reconstruction at CMS has been studied with very early data in $\sqrt{s} = 0.9$ TeV and $\sqrt{s} = 2.36$ TeV collisions [119], and in $\sqrt{s} = 7$ TeV [108] data, and a very good agreement with simulation was found, confirming the excellent modeling of the detector.

5.4 W and Z Signal Selection

The analyses presented in this thesis mostly rely on a sample as clean from background as possible, while the statistical error plays a minor role. Hence, emphasis is put on a large background rejection rather than high selection efficiency. Moreover, for both bosons a similar selection strategy is chosen in order to facilitate a good control of the systematic uncertainties. This applies especially for the relative measurements of $W^{\pm}$ and Z p_T. Particularly, it means the same muon selection and trigger criteria, jet selection, and, for the diffractive measurements, the same single vertex selection cuts.

5.4.1 $Z \to \mu^+\mu^-$ Signal Selection

The selection of the Z boson is rather straight-forward and based on the reconstruction of two isolated and high-energetic muons with an invariant mass peak close to the Z mass. The selection cuts for the two muons are those established in chapter 4. The event is thus selected online via the HLT trigger with a p_T threshold of 9 GeV (17 GeV) for 2010 (2011) data. Offline, the event is selected if exactly two muons satisfy the selection criteria with an additional cut on the transverse momentum of $p_T > 25\,\text{GeV}$.

The main source of background, due to its abundant production at the LHC, comes from hadronic decays of heavy and light flavor hadrons. These are mainly pions and kaons (light flavor), and $b\bar{b}$ and $c\bar{c}$ hadrons (heavy flavor), shortly called QCD events. They are usually characterized by low momentum and are therefore not isolated. The requirement of the isolation and the $p_T > 25\,\text{GeV}$ cut, successfully reject most of the QCD background.

Other sources for background are $t\bar{t}$ and other electroweak processes. The electroweak background has very similar kinematics and is called "irreducible". The main contribution comes from low cross-section processes such as the diboson events WZ and ZZ. Minor contributions come from WW and $W \rightarrow l\nu$ events where an additional jet is misidentified as a muon.

Thus, the most important backgrounds are reduced to an almost negligible contribution by the muon selection cuts. The remaining challenge for the Z selection is in the definition of the on-shell Z production and treatment of the non resonant Drell-Yan background. The 2 muons are selected if their invariant mass lies within $\pm 3\Gamma$ ($\Gamma = 2.5\,\text{GeV}$) of the nominal Z mass ($M_Z = 91.2\,\text{GeV}$). This choice is motivated by the ability to compare with theory calculations.

Table 5.1 shows the number of events, in successive order of applied selection cut, for the signal and the different backgrounds, as expected from simulation, together with the corresponding signal to background ratio. What should be noted in the line with the η cut, where the S/B decreases compared to the line before, is that this is rather a constraint from the detector geometry than a background rejection cut. It aims for a clear definition of the detector boundary and thus the S/B can evidently become smaller rather than larger. The remaining background after all cuts are applied is smaller than 0.2% and can be neglected.

Table 5.2 shows the number of selected events with all Z selection cuts applied for events with one, two or more jets, for signal and background as expected from simulation. The jet selection is the one specified in section 5.2.

Table 5.1: Number of events as expected from simulation, in successive order of applied selection cuts, for the Z signal and backgrounds per pb^{-1}, together with the signal to background ratio S/B. The MC samples are those detailed in section 2.3.

Sample	Z → $\mu\mu$	W → $l\nu$	QCD	$t\bar{t}$	VV	S/B
Produced	2'321	27'770	84'679	95	38	0.02
≥ 2 global μ	401	27	3'407	8.2	0.9	0.1
Triggered	376	18	195	4.2	0.7	1.7
Quality cuts	366	15	184	4.0	0.7	1.8
Isolated	340	1.2	7.3	0.9	0.5	34
$|\eta| < 2.1$	284	1.0	6.2	0.9	0.5	33
$p_T > 25$ GeV	234	0.01	0.01	0.5	0.3	268
Exactly 2 μ	234	0.01	0.01	0.5	0.3	273
$|M_{\mu\mu} - M_Z| < \pm 3\Gamma$	202	0.00	0.00	0.1	0.2	738

Table 5.2: Number of events as expected from simulation, for Z signal and background with one, two or more jets per 100 pb^{-1}, together with the signal to background ratio S/B. The MC samples are those detailed in section 2.3.

Sample	Z → $\mu\mu$	W → $l\nu$	QCD	$t\bar{t}$	VV	S/B
Z + 0 jets	17'080	0.2	0.0	0.2	6.5	2'480
Z + 1 jet	2'486	0.1	0.0	1.5	8.3	253
Z + 2 jets	521	0.0	0.0	2.8	4.6	70
Z + 3 or more jets	112	0.0	0.0	1.9	1.4	35

5.4.2 Results with $\sqrt{s} = 7$ TeV Data

The muon selection variables which were studied using MC simulation were measured with the data collected from very different periods of average instantaneous luminosity and pileup conditions. With increasing collected data the advanced understanding of the detector could be implemented in the simulations. The results from the 2010 and 2011 data taking periods are compatible and thus an extract of the 2011 data is shown here. Figure 5.4 shows the muon selection variables for a sample of data collected in 2011 with an integrated luminosity of 1.5 fb^{-1}. The variables are shown for events with all Z selection cuts applied, except for the one shown in the figure, which is applied only for one of the two muons. As was already noted in table 5.1, the signal is almost background free and the latter can be neglected. An excellent agreement between data and MC is found for all selection variables.

Figure 5.5 shows the azimuthal angle ϕ and the pseudorapidity η for the selected events. The data are in very good agreement with MC for both distributions. Noticeable is the

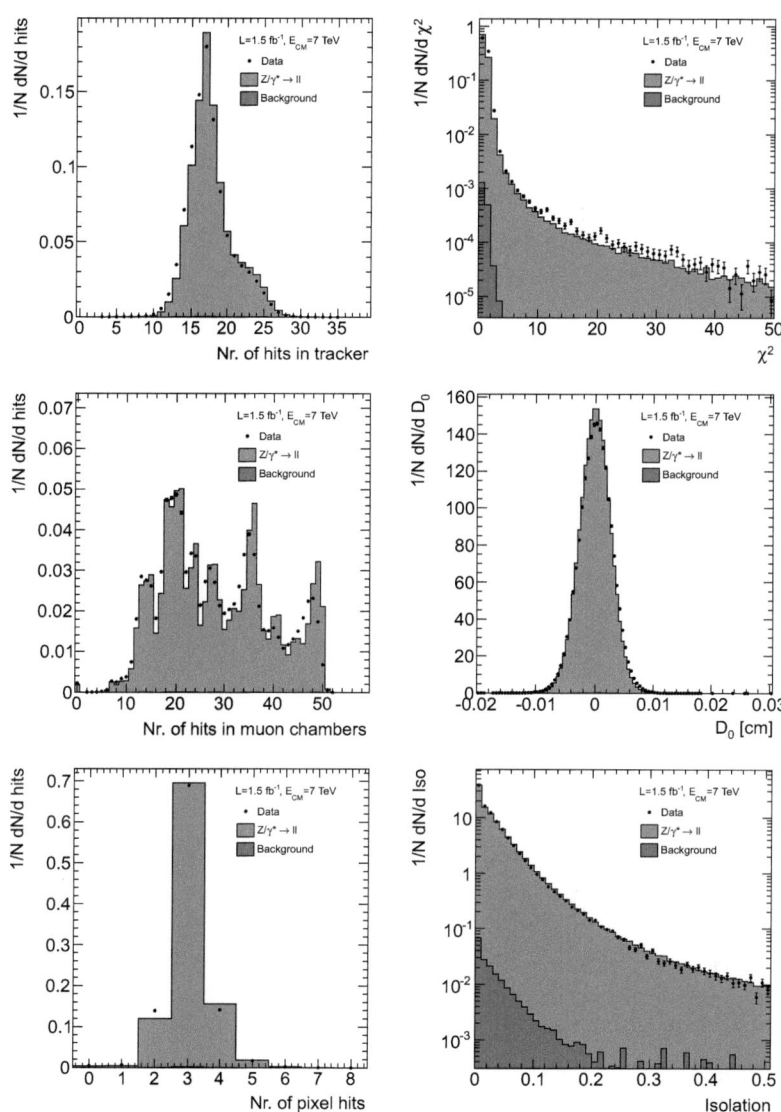

Figure 5.4: Muon selection variables for $1.5\,\text{fb}^{-1}$ data collected in 2011 compared to signal and background simulations. All Z selection cuts are applied except the one shown in the figure, which is applied only on one of the muons.

good agreement in the η distribution in the regions of the dips, where the wheels of the muon system overlap (cf. 4.1). This agreement was achieved only with increased collected data and was not yet given in the low luminosity period.

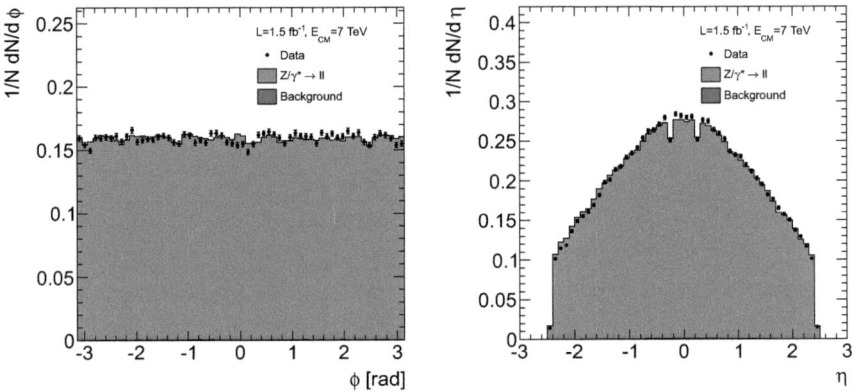

Figure 5.5: Muon azimuthal angle ϕ (left) and pseudorapidity η (right) distributions for the selected Z events for data and simulation and for an integrated luminosity of $1.5\,\text{fb}^{-1}$.

Figure 5.6 shows the transverse momentum spectrum of the two muons composing the Z, with the higher and the lower p_T for data and MC in linear and logarithmic scale. The agreement is excellent over the entire range, only a small shift is observable in the spectrum for the muon with the lower p_T.

Finally, in fig. 5.7 the invariant mass distribution of the selected dimuon system is shown for all accepted events in linear and logarithmic scale including the ratio of data over MC. The MC distributions are normalized to data with their corresponding cross-sections (cf. section 2.3). A very small shift of data to the left can be observed compared to MC. The ratio visualizes this fact more clearly and it can be seen that the effect is very small. This shift is explained by the residual momentum scale misalignment of the tracker as was shown in fig. 5.6 and directly affects the invariant mass distribution. Nevertheless, this effect is very small, comparing for example with the dielectron system where the shift coming from the crystal deterioration is more than a factor of 5 larger, and can be neglected for the following measurements in this thesis.

The invariant mass distribution is also shown for selected Z events where no jet, exactly one or exactly two jets are reconstructed, with the jet complying the definition in section 5.2 (fig. 5.8). The slight shift which is observed in the inclusive case is also visible in the different jet bins. Otherwise the data distributions agree well with the simulations and no particular feature can be observed in any of the jet cases.

All variables and distributions presented in this section point out how well the detector is understood and how the simulations are able to reproduce in a very good approximation the measured data. This is relevant and required for any further studies of Z variables which will be presented in the subsequent chapters.

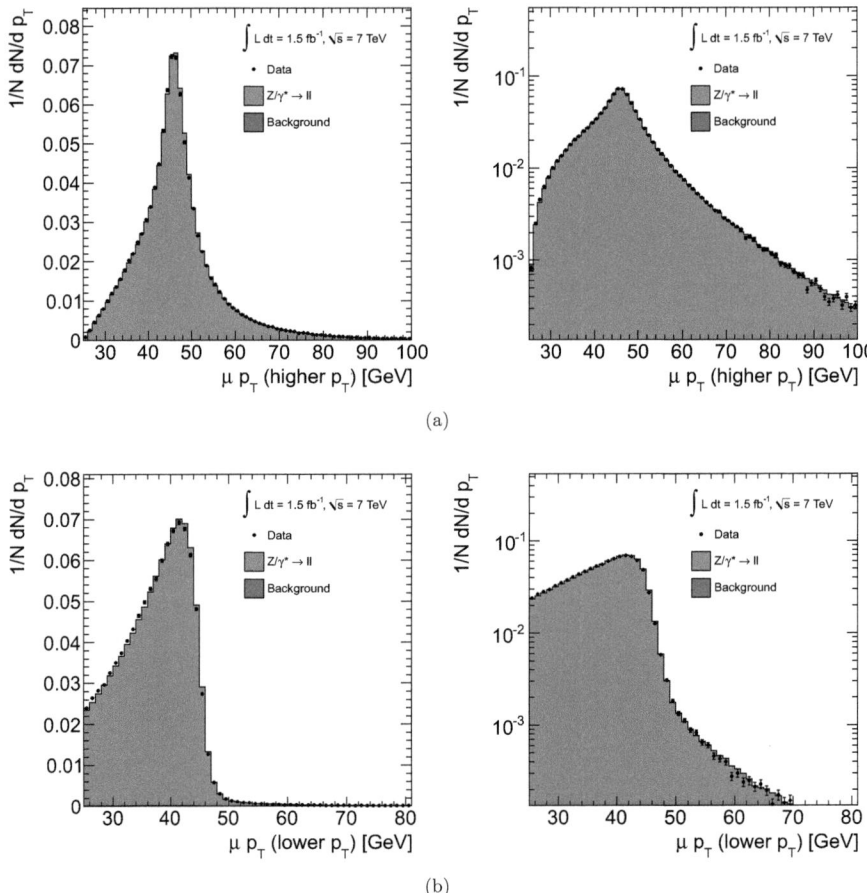

Figure 5.6: Lower and higher transverse momentum distributions for the two muons composing the selected Z events for data and simulation and for an integrated luminosity of $1.5\,\mathrm{fb}^{-1}$. The muon with the higher p_T is shown in (a), while (b) shows the lower p_T distribution. Linear scale is on the left, logarithmic on the right.

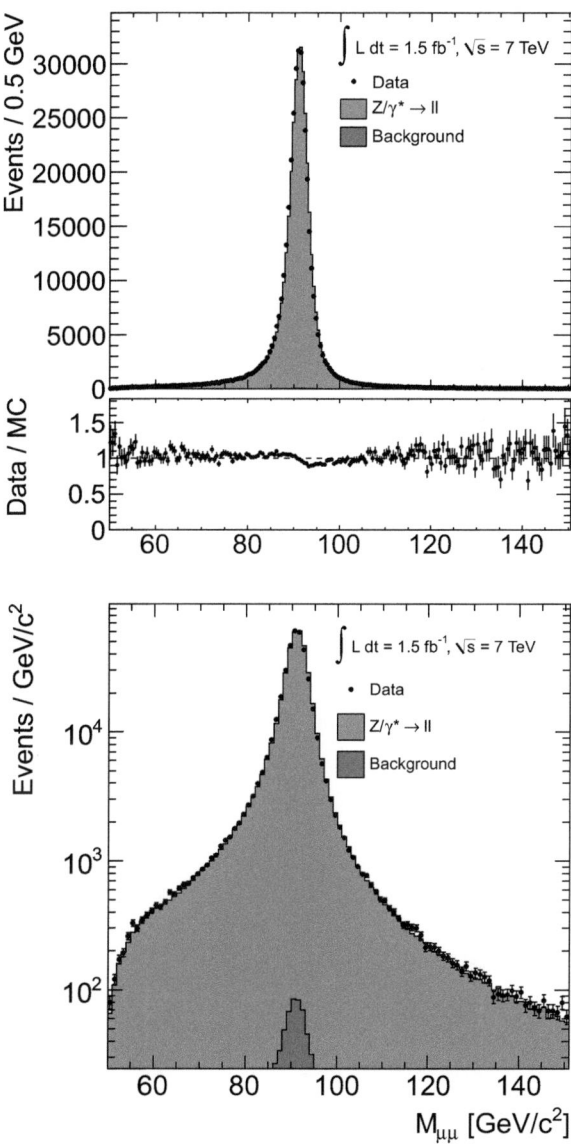

Figure 5.7: Invariant mass distribution for the dimuon system for all accepted events after the Z selection cuts in linear and logarithmic scale in data and MC including the ratio of data over MC.

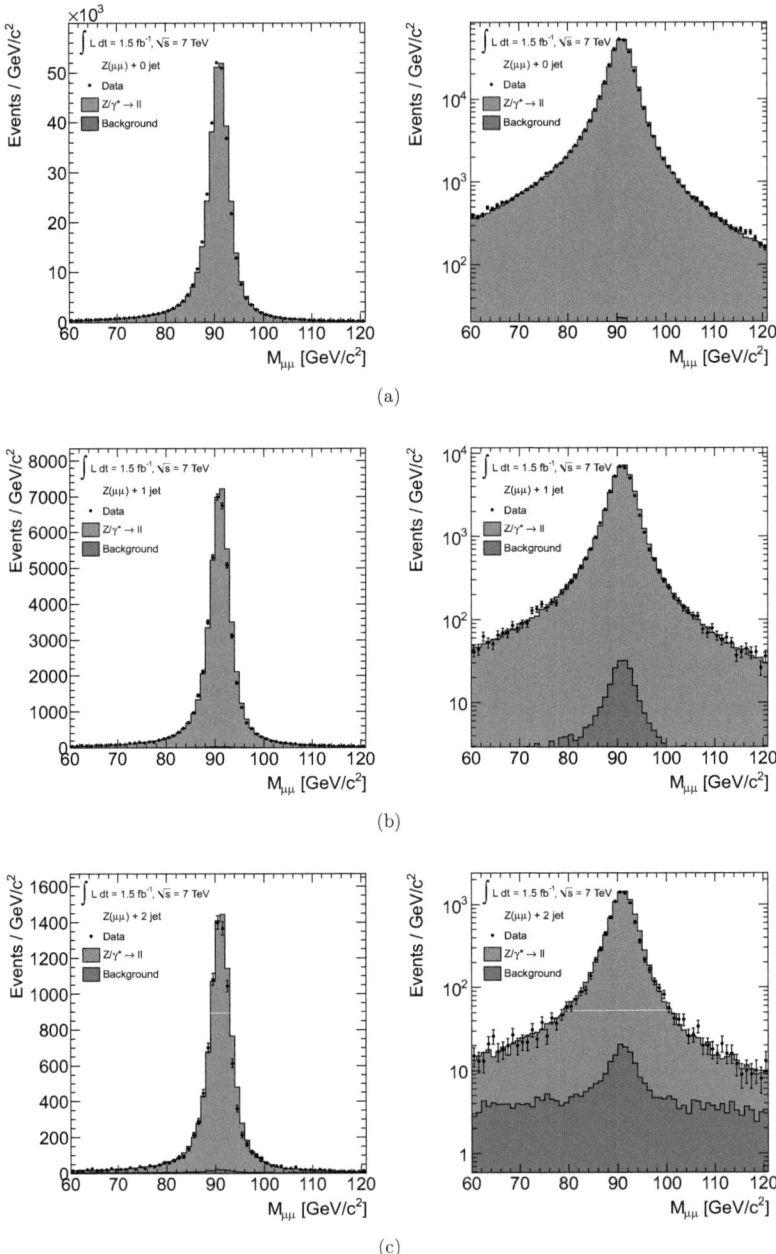

Figure 5.8: Dimuon invariant mass for selected Z events with (a) no, (b) one, and (c) two jets for data and MC in linear (left) and logarithmic (right) scale. The jet is defined as in section 5.2.

5.4.3 W → μν Signal Selection

Due to the neutrino, which is measured through the large missing E_T, only the transverse mass of the W

$$m_T^W = \sqrt{2 \cdot p_T^\mu \cdot p_T^\nu \left(1 - \cos\left(\Delta\phi(\mu,\nu)\right)\right)}$$

can be determined. Thus, the selection of the W with subsequent leptonic decay, is not as straight forward as the Z selection. Compared to the very accurate p_T measurement of the muon, the MET reconstruction is much less precise. In spite of this, a good signal to background ratio can be obtained by applying somewhat tighter cuts. A W event candidate is selected if it contains exactly one muon satisfying all selection cuts, specified in chapter 4, and a $p_T > 25\,\text{GeV}$. Additionally, the missing E_T has to be greater than 30 GeV. To reduce the Drell-Yan background, candidates are rejected if they have two muons satisfying the selection cuts, where one of them has a $p_T > 20\,\text{GeV}$ and the second muon a $p_T > 10\,\text{GeV}$. Finally, a transverse mass cut of $\text{M}_T > 60\,\text{GeV}$ is applied.

The number of events is shown for each step, after applying the cut, in table 5.3, for the W signal and background events per pb^{-1}, as expected from simulation and together with the corresponding S/B.

Table 5.3: Number of events, in successive order of applied selection cuts, for the W signal and backgrounds, as expected from simulation per pb^{-1}, together with the signal to background ratio S/B. The MC samples are those detailed in section 2.3.

Sample	W → μν	Z → ll	QCD	$t\bar{t}$	VV	S/B
≥ 1 global μ	7'494	836	78'980	40	8	0.094
Triggered	4'998	612	4'588	14	4	0.958
Quality cuts	4'952	609	4'512	14	4	0.963
Isolated	4'882	605	2'050	14	4	1.83
$|\eta| < 2.1$	4'442	569	1'905	13	4	1.78
$p_T > 25\,\text{GeV}$	3'527	477	210	11	3	5.0
Exactly 1 μ	3'527	243	210	11	3	7.5
Z rejection	3'527	207	210	11	3	8.2
MET > 30 GeV	2'413	65	12	8	2	27.4
$\text{M}_T > 60\,\text{GeV}$	2'242	55	1	6	2	35.3

Similarly to the Z selection, the largest source of background, due the huge cross-section, are the QCD events. The isolation and p_T criteria (already included to a certain extent in the trigger), are able to reject most of them, but some contamination remains. After applying the MET cut, they are reduced to a very small fraction of the total background.

The M_T cut reduces them even further, making the QCD events the smallest background contribution.

The contamination from $Z \to \mu\mu$ events, where one of the muons is not reconstructed due to detector gaps or quality reasons and is thus measured as missing E_T, are efficiently reduced by the Z rejection cut and overall the S/B increases significantly.

The contribution of $t\bar{t}$ and diboson backgrounds are small to negligible in the inclusive sample. Nevertheless, as can be seen in table 5.4, the $t\bar{t}$ background becomes considerable with increasing jet activity. The favored top decay mode is the decay into a W and a b-quark. The resulting signature can therefore be a muon from one of the Ws plus several jets. This is especially true in the three (or more) jet case, as can be seen from table 5.4. This background can be reduced by a veto of more than 2 jets, if those W + jet bins are not being studied. In this thesis, this cut is not applied as it is very small in any case.

Table 5.4: Number of events as expected from simulation for W signal and background with one, two or more jets per 100 pb^{-1} together with the signal to background ratio S/B. The MC samples are those detailed in section 2.3.

Sample	$W \to \mu\nu$	$Z \to ll$	QCD	$t\bar{t}$	VV	S/B
W + 0 jets	196'535	4'854	62	3	51	39.6
W + 1 jet	22'024	504	35	32	64	34.6
W + 2 jets	4'628	119	12	123	37	15.9
W + 3 or more jets	1'043	35	4	400	11	2.3

To summarize, the S/B ratio for the W selection is much smaller than for the Z events, which is expected, but it is still very high and allows for precise measurements of W variables.

5.4.4 Results with $\sqrt{s} = 7$ TeV Data

As for the Z events, only a selection of 2011 data is shown, as the results are compatible for the different data taking periods. To exclude any correlation in the muon selection variables shown for the Z events, where one of the muons always had to fulfill all the conditions, including the one shown for the second muon, here all the selection variables are shown for the muon composing the W candidate fulfilling the W selection requirements (fig. 5.9).

The results are almost identical to the Z events. The only variable which shows a slight deviation, is the isolation in the tail, where the statistical uncertainty is large. Due to

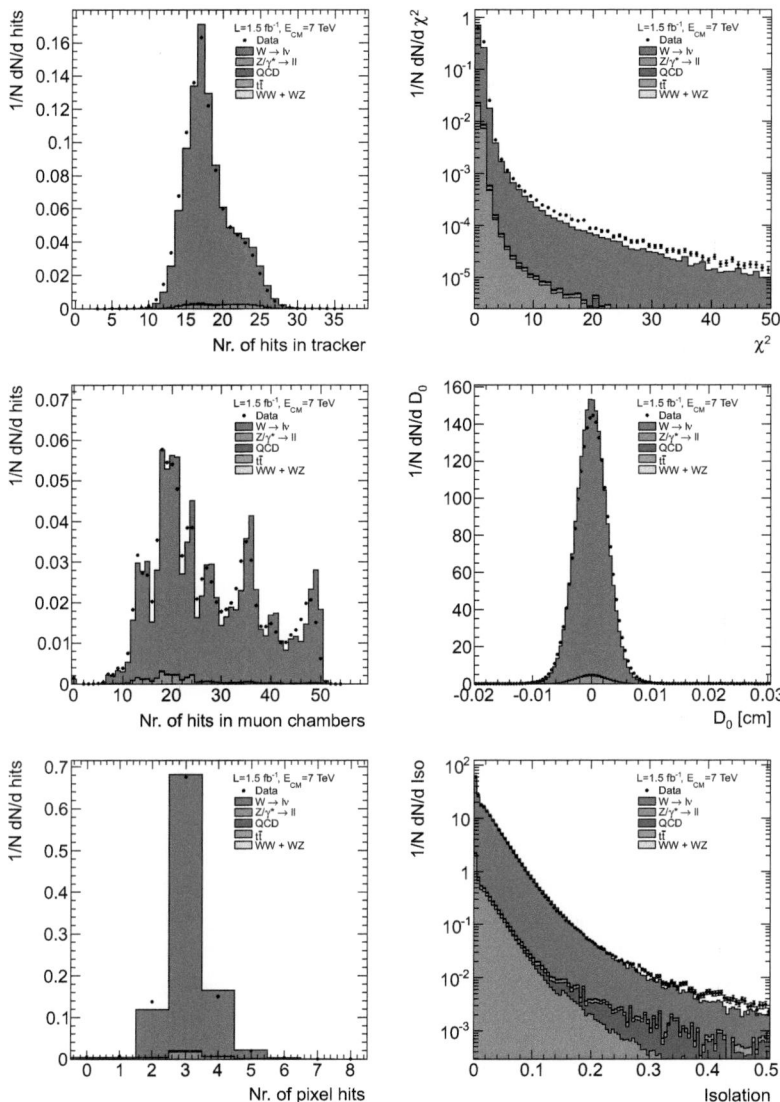

Figure 5.9: Muon selection variables for $1.5\,\text{fb}^{-1}$ data collected in 2011 for W candidate events compared to signal and background simulations. All W selection cuts are applied except the one shown in the figure.

the cut applied at 0.15, which is safely away from the discrepancy, this does not affect the selection. The kinematic distributions ϕ and η are shown in fig. 5.10. The agreement of data with MC is reasonable for the pseudorapidity distribution while the ϕ variable shows a slightly worse agreement than in the Z events.

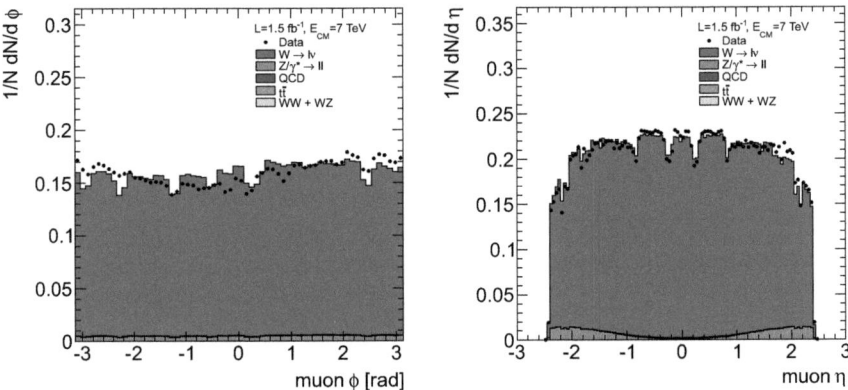

Figure 5.10: Azimuthal angle ϕ (left) and pseudorapidity η (right) distributions for the selected W events for data and simulation and for an integrated luminosity of $1.5\,\text{fb}^{-1}$.

The transverse momentum of the selected muon and the missing E_T distribution are shown in linear and logarithmic scale before applying the M_T cut (fig. 5.11) and after (fig. 5.12). The agreement for the muon p_T is already good before the M_T cut. Only a slight broadening of the data spectrum compared to simulation can be observed. The MET distribution on the other hand does not agree perfectly in fig. 5.11, a broadening of the peak is observed and a deviation of the shape between roughly 10-20 GeV. After applying the M_T cut, which is directly correlated to the measured p_T of the muon and the neutrino, the distribution shows a very good agreement apart from a minor broadening of the shape. The QCD background, which is the least understood background process and therefore introduces the largest shape distortions at low energies, has disappeared almost completely. For this reason, it can safely be assumed that all the distributions reproduce data accurately enough in the relevant phase-space and we can proceed to study W properties.

In fig. 5.13 the transverse mass distribution is shown for the muon-MET system for all selected W events in linear and logarithmic scale for data and simulation and the ratio of data to MC is given. The MC distributions are normalized to data with their corresponding cross-sections (cf. section 2.3). MC reproduces data very well; a small broadening of data compared to simulation can be observed, consistent with the previous findings.

Finally, the M_T distributions for W events with no, exactly 1, or exactly 2 jets is shown in fig. 5.14, with the jets defined as in section 5.2. Some broadening similar to the inclusive case can be observed in all three figures. What can be noted from the comparison of the three event classes, is the increasing $t\bar{t}$ background with increasing jet activity, which was already detailed in the previous section.

5.4.5 $Z \to e^+e^-$ and $W \to e\nu$ Signal Selection

For completeness, the W and Z selection in the electron decay channel is detailed here, as it will be used to present the combined measurement in the events with a large pseudorapidity gap (chapter 8). The selections in the two channels are chosen to be as similar as possible. This particularly means the same jet and MET reconstruction algorithm and cuts, vertex selection, invariant and transverse mass cuts, lepton transverse momentum cut and acceptance restriction. The only difference is in the lepton identification, due to the different detectors used in the reconstruction. An electron has to fulfill the following requirements to be accepted:

- $|\Delta\eta|$ between extrapolated track and calorimeter supercluster has to be smaller than 0.004 (0.007) in the barrel (endcap)
- $|\Delta\phi|$ between extrapolated track and ECAL supercluster has to be smaller than 0.06 (0.03) in the barrel (endcap)
- $H/E < 0.04$ (0.025), where H is the hadronic energy and E the electromagnetic energy measured in the calorimeters
- $\sigma_{i\eta i\eta} < 0.01$ (0.03) (this is a shower shape variable) in the barrel (endcap)
- combined isolation variable $E_{\text{Iso}}^{\text{comb.}}/E_{\text{electron}} < 0.07$ (0.06) to select isolated electrons only; combined because the tracker and calorimeters are combined
- electrons from conversions are removed

For more details on electron reconstruction and performance see [83].

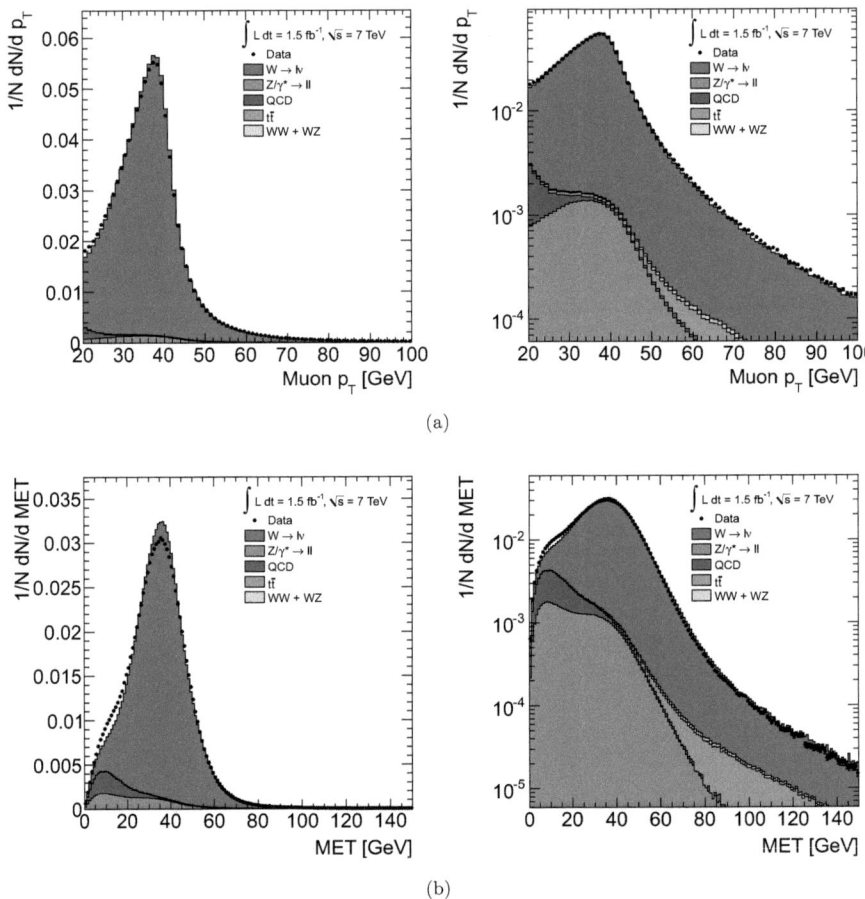

Figure 5.11: Muon transverse momentum (a) and missing E_T (b) distributions for the selected W events before applying the M_T cut, for data and simulation and for an integrated luminosity of $1.5\,\text{fb}^{-1}$ in linear (left) and logarithmic (right) scale.

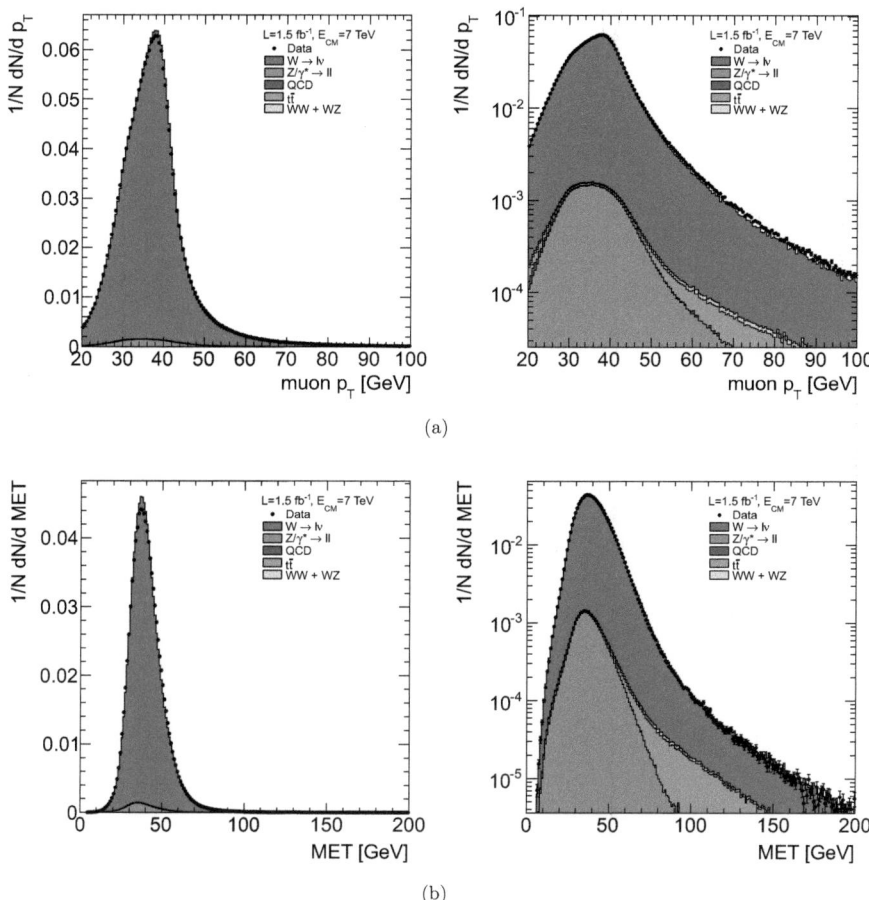

Figure 5.12: Transverse momentum (a) and missing E_T (b) distributions for the selected W events after applying the M_T cut, for data and simulation and for an integrated luminosity of $1.5\,\text{fb}^{-1}$ in linear (left) and logarithmic (right) scale.

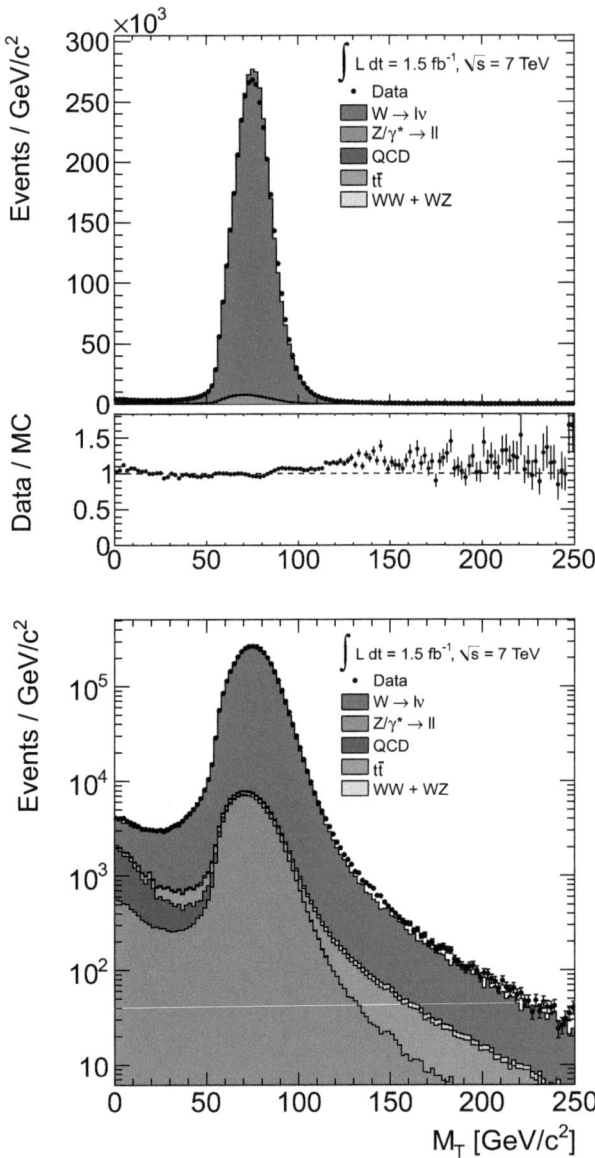

Figure 5.13: Transverse mass distribution for the muon-MET system for all accepted W events in linear and logarithmic scale in data and MC including the ratio of data over MC.

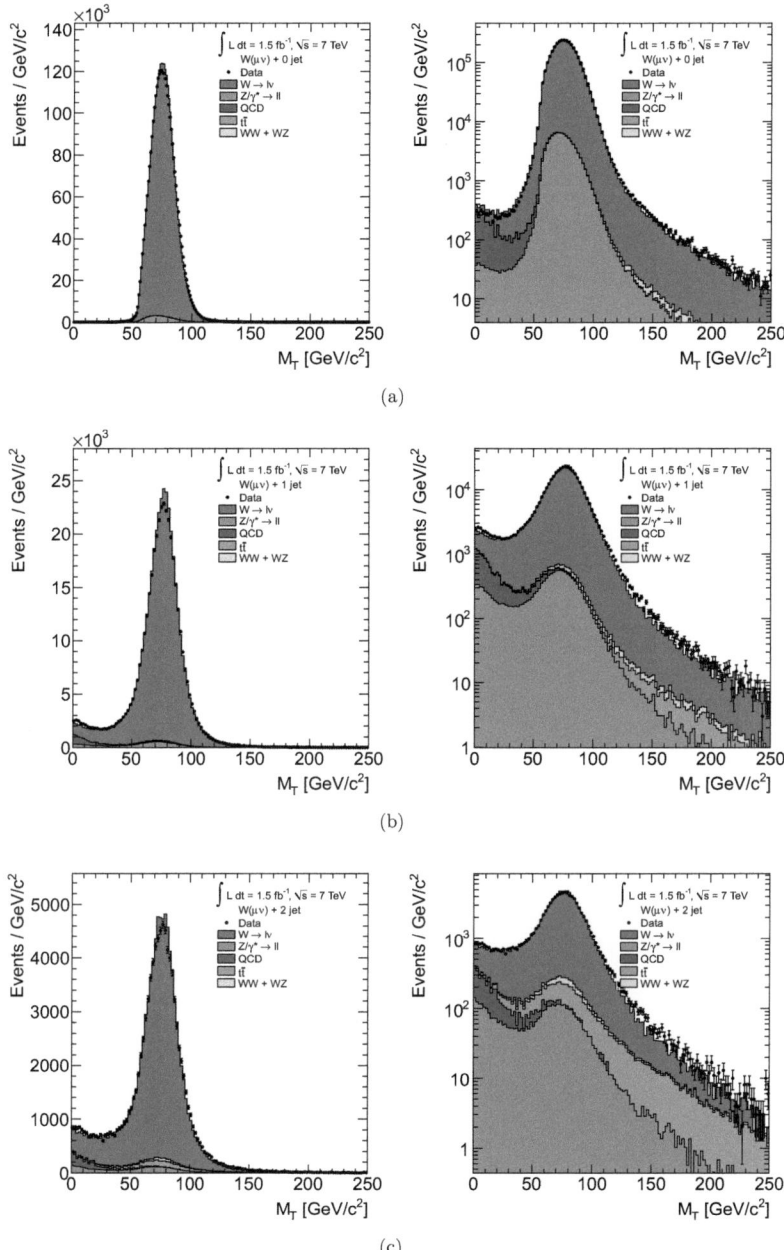

Figure 5.14: Transverse mass distribution for the muon-MET system for selected W events with (a) no, (b) one, and (c) two jets for data and MC in linear (left) and logarithmic (right) scale. The jet is defined as in section 5.2.

Chapter 6

Diffractive processes in W and Z events

The study of hard-diffractive event observables, such as a large rapidity gap in W or Z boson events, is strongly influenced by the underlying multiparton interactions. At present, the description of these processes relies largely on the generator specific tuning of data. Therefore, the measurement and extensive study of such events can be used in the understanding and implementation of detailed simulations of diffraction. Moreover, the measurements of single pomeron exchange are very useful due to their sensitivity to the structure of the pomeron and the dynamics of diffraction. In this and the two following chapters, a detailed overview of the analysis steps towards the observation of a diffractive signal, with improvements of Monte Carlo simulations containing multiparton interactions is given. Central charged-particle multiplicities, forward energy flow, and correlations between them will be presented in W and Z events, identified by the vector-boson decays to muons and electrons, using the 2010 data sample of pp collisions at 7 TeV, corresponding to an integrated luminosity of 36 pb^{-1}. In this work the decay to muons is emphasized. Consistent results are obtained with the electrons, and some of the results are presented in combination, especially when the result is more meaningful, e.g. due to statistical errors dominating the results precision, or to emphasize the consistency. The complete electron results can be found in the publication[1].

6.1 Event Topology

Diffractive events can be selected through the detection of a large rapidity gap in the event or by measuring the non-dissociated proton. The latter needs very specific detectors such as the by now standard roman pots used at HERA, Tevatron, and also at the TOTEM experiment at the LHC, which are located at tens or hundreds of meters away from the interaction point, and which can detect leading protons scattered at very small angles. CMS has no roman pots and the first method had to be applied. The forward hadronic calorimeter (HF), which can measure particle energies up to a pseudorapidity of 5.2, were used. This method is only suitable in the low luminosity data taking period due to a large number of pileup events at high luminosities, where the gap is filled with particles coming from a different interaction. The effects of the pileup events on the forward energy flow measurements will be discussed in the next section. Figure 6.1 illustrates the gap signature with an event display graph of CMS data. Two different perspectives of the detector are shown for a $W \to \mu\nu$ candidate with a large pseudorapidity gap. It visualizes the hit of the muon in the muon chambers and the missing energy represented by an arrow. The empty detector region on the right part indicates the large pseudorapidity gap. For comparison an event display with a $Z \to \mu\mu$ event without LRG is shown in fig. A.1 and one with a gap in fig. A.2 in the appendix.

However, the LRG signature is not unambiguous because a gap can also be the result from a fluctuation in an inelastic event. The measured LRG events will thus be an overlap of real pomeron exchange events and those resulting from fluctuations in the hadronisation process.

Additionally to the hard interaction, the multiparton interactions will leave traces in the detector. Thus, the measured energy and particle multiplicity distributions will be an overlap of the two. Both, the hard process and the multiparton interactions can be produced diffractively. Correspondingly, different correlations between central and forward distributions are expected, depending on the process. This is pictured in fig. 6.2. The standard picture of W or Z boson production via hard parton-parton scattering is shown in fig. 6.2a and combined with a multiparton interaction in fig. 6.2b. According to this, large rapidity gap events can only arise from multiplicity fluctuations. Figure 6.2c shows standard W(Z) production accompanied by multiparton interactions with a diffractive component. Hard-diffractive production (fig. 6.2d) leads to a large rapidity gap. Such contributions would result in almost unchanged central charged-particle multiplicity distributions with respect to fig. 6.2b. A larger fraction of events with a relatively small energy deposition in the forward regions, and a smaller correlation of the energy flow in the central and forward regions, could be expected. The diffractive production mechanism with multiparton interactions is shown in fig. 6.2e. In this case, large rapidity gap events only survive if the multiparton component is small. Finally, fig. 6.2f indicates a possible combination of W(Z) production mechanisms with a diffractive component in both the hard process and the multiparton interaction.

6.2 Pileup Influence and Rejection

The identification of W and Z events is the one given in chapter 5. From this starting point, further selection criteria are chosen. It is important to note, that the analysis described in the following aims for a relative measurement of W and Z production with a LRG with respect to the inclusive production rate.

As mentioned earlier, there can be several simultaneous pp interactions in the same bunch crossing in addition to the selected W and Z events. As the number of bunches and the instantaneous luminosity increased steadily during the 2010 pp data taking, the analysis was affected by very different pileup conditions. For this analysis the data have been separated into three periods (P) with average instantaneous luminosities of $L_{inst} \leq 0.17$ μb^{-1}/s (P I), $0.17 < L_{inst} \leq 0.34$ μb^{-1}/s (P II), and $L_{inst} > 0.34$ μb^{-1}/s (P III). Figure 6.3 shows the number of reconstructed vertices in the W $\rightarrow \mu\nu$ sample for the three periods. Assuming a total inelastic cross section of about 70 mb [120, 121], an instantaneous luminosity of 0.17 μb^{-1}/s (P I) corresponds to an average of about one inelastic pileup event,

$$N = \frac{\dot{L}\sigma}{f} \sim 1,$$

with $\dot{L}$ the average instantaneous luminosity, σ the total inelastic cross section and f the revolution frequency. Accordingly, P II corresponds to an average of 1–2 and P III to about 3 pileup events.

The selection efficiency for W and Z events is independent of the instantaneous luminosity. However, the charged-particle multiplicities and the energy depositions in the forward region of the detector, and thus especially the LRG signature, are strongly affected by the pileup (i.e., the gap is filled in). In order to limit the consequences of pileup, events with more than one vertex are rejected. For this analysis a primary vertex, the W(Z)-vertex, is defined as the one which contains the lepton track(s). Events with additional vertices, formed by at least three tracks, are rejected. The W-vertex z-position distribution is

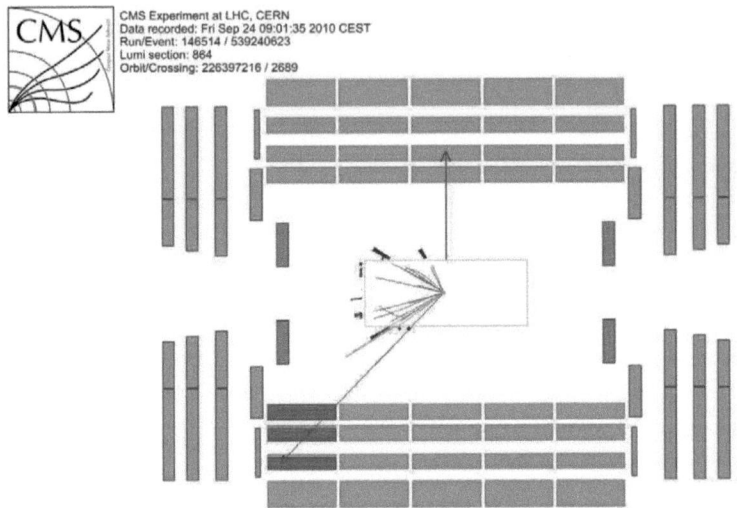

(a) r-z view with the positive pseudorapidity on the right side

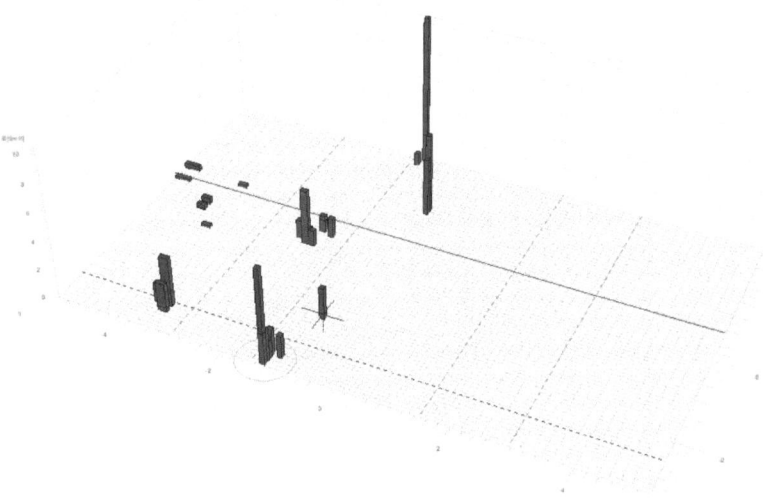

(b) Lego plot view of the detector

Figure 6.1: Event Displays for a $W \to \mu\nu$ candidate with a large pseudorapidity gap on the positive detector hemisphere for two different views visualizing the particle deposits with a large gap in the detector.

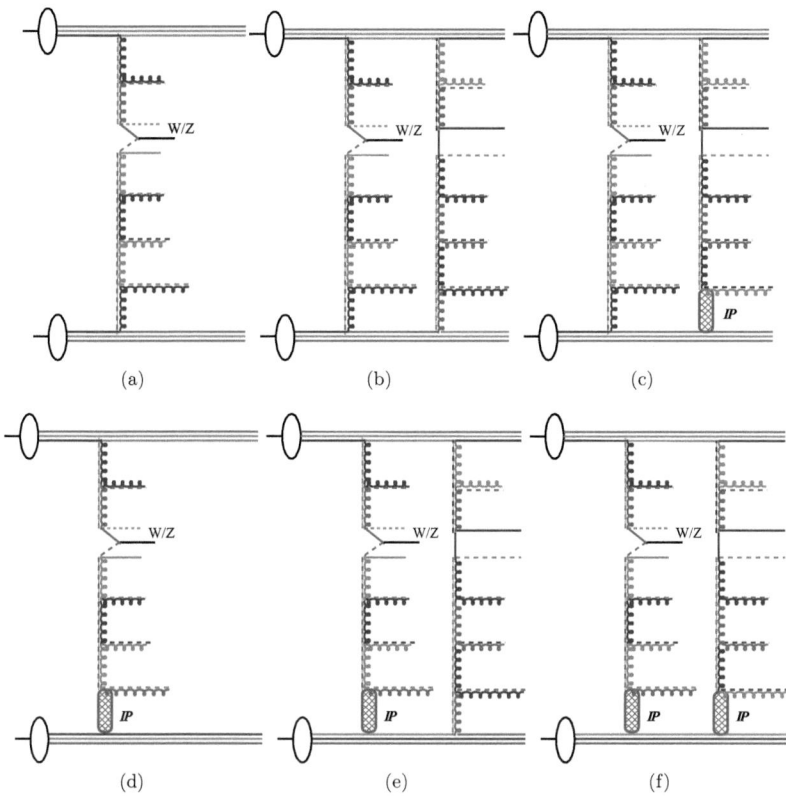

Figure 6.2: Sketches of the process pp → W(Z)X, where a W(Z) boson is produced in the hard interaction, combined with contributions from multiparton interactions, with and without a diffractive component. The colored curly and straight lines are for gluons and quarks and the straight black lines are for W(Z). The symbol $I\!P$ indicates the exchange of a state with the quantum numbers of the vacuum (Pomeron). (a) shows the standard hard interaction; (b) the same process with additional multiparton interactions; (c) the hard process accompanied by multiparton interactions containing a diffractive component; (d) the hard-diffractive production of a W(Z) boson; (e) the hard-diffractive W(Z) production with multiparton interactions, and (f) the hard-diffractive W(Z) production with multiparton interactions containing a diffractive component. In the latter case, the diffractive component of the multiparton interaction does not necessarily couple to the same proton as in the hard process (not shown here). [1]

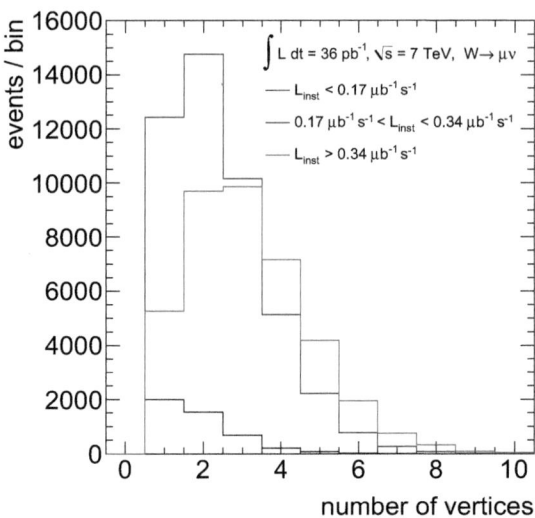

Figure 6.3: Number of reconstructed vertices for three data-taking periods of different average instantaneous luminosities.

roughly Gaussian with a mean of 0.52 cm and a standard deviation of 5.9 cm.

The effects from pileup events on the analysis have been studied by means of zero-bias data samples, where the only requirement was that of beams crossing in the detector. The event samples were analyzed for the three different instantaneous luminosities periods. Pileup events can be categorized in hard and soft events. Hard pileup interactions have some detectable charged particles in the central region of the detector and are removed by the multiple-vertex veto. The soft component has little or no detectable transverse activity in the central region and does not result in reconstructed vertices.

In fig. 6.4 the MC efficiency to reconstruct a pileup vertex as a function of the true distance in z from the W-vertex is shown for events with exactly one pileup, including soft pileup in $WX \to \mu\nu X$ events. The three lines correspond to three different vertex quality cuts regarding the number of tracks which form the vertex (cf. section 5.3). Based on the pileup conditions in the 2010 data, the efficiency to detect pp pileup interactions was found to be essentially constant within ± 25 cm along the z direction of the nominal interaction point with an average of about 72%. The inefficiency essentially depends only on the amount of soft pileup interactions (e.g., events without detectable charged particles in the central region of the detector) in the MC simulation. The reconstruction efficiency for pileup vertices was found to be essentially independent for the different luminosity periods, as long as the vertices were separated by more than 0.1 cm. This is illustrated in fig. 6.4 where no vertex is found when it is too close to the W-vertex. The corresponding inefficiency to detect the merged pileup vertex is estimated from a fit to the dip to 3.3%.

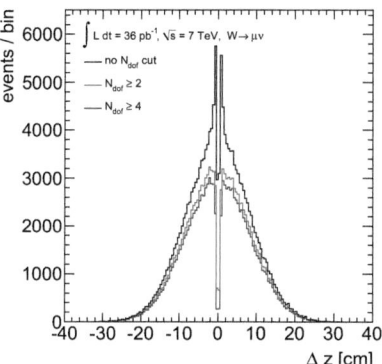
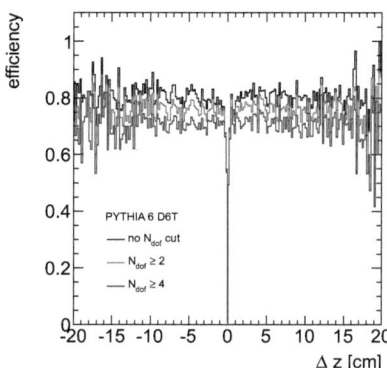

Figure 6.4: Distance between the reconstructed W-vertex and an additional vertex, illustrating the inefficiency to reconstruct a vertex if the distance between them is less than 0.1 cm (left) and efficiency to reconstruct a pileup vertex as a function of the true distance in z from the W-vertex with exactly one pileup in $WX \to \mu\nu X$ events, again illustrating the inefficiency around zero (right). The three lines correspond to three different cuts of vertex quality regarding the number of tracks which form the vertex (cf. section 5.3).

The numbers of single-vertex W and Z events are summarized in table 6.1. The single-vertex event yields, compared to the inclusive event yields, decrease with increasing instantaneous luminosity and are in agreement with the expected numbers of vertices from the simulation, assuming Poisson distributions.

Table 6.1: Number of W and Z candidate events with a single primary vertex. The numbers are given for the total and the three data-taking periods of different instantaneous luminosities. The percentage of single-vertex events with respect to all selected W and Z candidates for that period is given in parentheses.

Single-vertex events	$W \to \mu\nu$	$Z \to \mu\mu$
Total	17924 (26.2%)	2924 (26.1%)
P I	1926 (53.0%)	328 (56.8%)
P II	10524 (32.7%)	1718 (31.0%)
P III	5474 (17.3%)	878 (17.2%)

Chapter 7

Forward Energy Flow and Central Charged-Particle Multiplicity

In this chapter, the observables used to study the underlying event structure in W and Z events, the charged-particle multiplicity in the central detector and the energy depositions in both forward calorimeters, are presented. Correlations between these distributions are analyzed and put in context with the simulation of diffractive W/Z events, with a diffractive component in either the hard process, or the underlying multiparton interactions, or both of them.

In the following, the forward calorimeters are designated as HF+ and HF−, depending on the sign of the corresponding η coverage. Only the distributions for $W \to \ell\nu$ events are discussed. This work concentrates on the muon channel but no significant differences are observed between W events selected with decays to electrons or muons. The same analysis has been performed on the pp $\to$ ZX data sample and consistent results are obtained. For completeness, the distributions corresponding to the Z data are appended at the end of this document.

7.1 Results for the Forward Energy and Central Multiplicity Measurements

The charged-particle multiplicity is measured in the range $|\eta| < 2.5$ for track momentum thresholds of $p_T > 0.5$ GeV and $p_T > 1.0$ GeV, excluding the track associated with the W decay. Additionally, in order to study the underlying event structure, events with central jet activity ($p_T > 30$ GeV, $|\eta| < 2.5$) are excluded from the multiplicity plots.

A detailed study of the pion track reconstruction within the acceptance of the tracker [118] determined that the efficiency rises from about 88% at a p_T of 0.5 GeV to about 95% for p_T between 1–10 GeV. Above 10 GeV, the efficiency decreases slowly to about 90% at 50 GeV. Furthermore, it was shown that the hadron track reconstruction efficiency in the data agrees within 1-2% with the one in the MC simulation. The total systematic uncertainty of the tracking efficiency was estimated to be less than 3.9%.

The observed charged-particle multiplicities, excluding the lepton(s) from the W(Z) decays, vary between 0 and about 50, with an average of 11 and an r.m.s. of 8.2 for $p_T > 1.0$ GeV. About twice as many tracks are found with the lower threshold, $p_T > 0.5$ GeV, and about 0.15% of the events have more than 100 tracks.

The charged-particle multiplicity distribution for W $\to \mu\nu$ events is shown in fig. 7.1a for tracks with $p_T > 0.5$ GeV, and in fig. 7.1b for $p_T > 1$ GeV. The PYTHIA8 generator with tune 2C provides the best overall description for the higher track p_T threshold, and PYTHIA6 with tune Z2 provides a reasonable description for both track p_T thresholds. However, both PYTHIA8 2C and PYTHIA6 Z2 predict too many events with very small charged-particle multiplicities. The PYTHIA6 D6T tune predicts a harder p_T spectrum for hadrons in the underlying event and thus a larger multiplicity in the case of the higher threshold. The Pro-Q20 tune of PYTHIA6 significantly underestimates the event yields with very high multiplicities.

In contrast to the multiplicity distributions, the charged-particle transverse momentum spectrum (fig. 7.1c) shows a very good agreement with all the MC tunes.

The energy deposition in the HF+ and HF− calorimeters is determined from the sum of individual calorimeter towers with an energy threshold of 4 GeV, corresponding to a minimum transverse momentum of 0.07–0.4 GeV. The uncertainty on the energy scale of the HF calorimeter was estimated to be about $\pm 10\%$ [122]. This uncertainty was taken into account by a $\pm 10\%$ scaling of the single-tower energy, resulting in new estimates of the total energy deposition in HF for the data, while keeping the MC unchanged. In fig. 7.2 and all the following figures, the data points are plotted in the center of the corresponding bins, and the corresponding systematic uncertainty for the energy measurement in the HF is shown as a band. This uncertainty is much larger than the 3.5% difference between the reconstructed energy distributions in the HF+ and HF− calorimeters (cf. table 7.1).

The observed HF energies vary between 0 GeV (i.e., no HF tower with an energy above 4 GeV) and more than 2 TeV. Only a few events have an energy deposition above 2 TeV

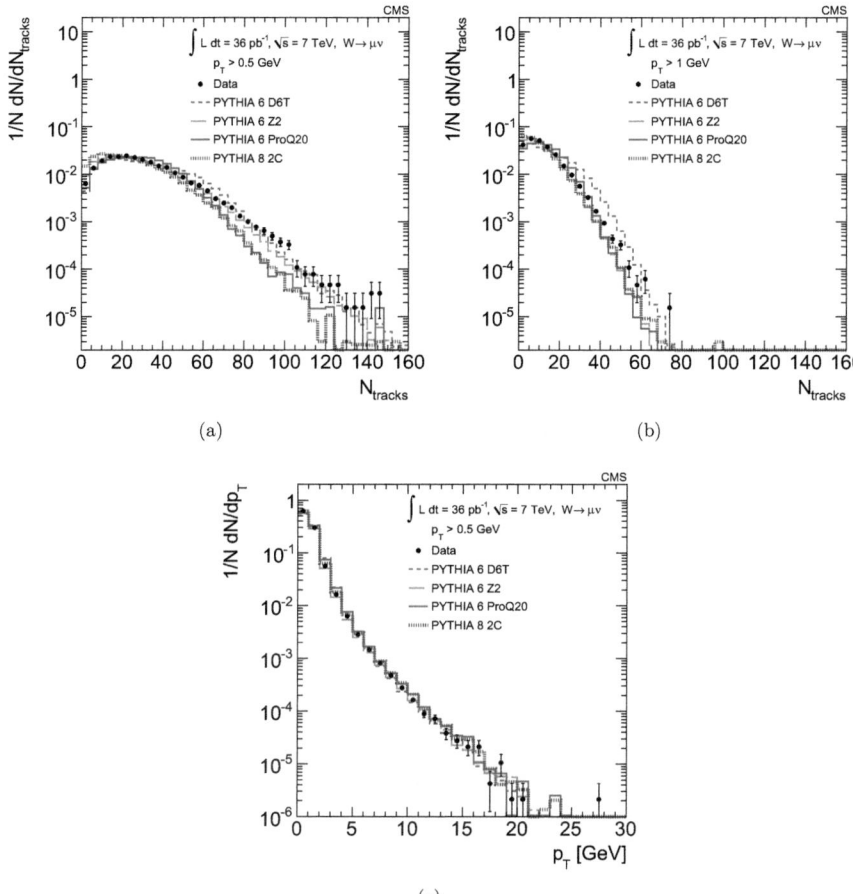

Figure 7.1: (a) and (b) charged-particle multiplicities for two different p_T thresholds and (c) transverse momentum spectrum for W $\to \mu\nu X$ candidate events shown for data and MC simulations with a selection of different tunes for the underlying event.

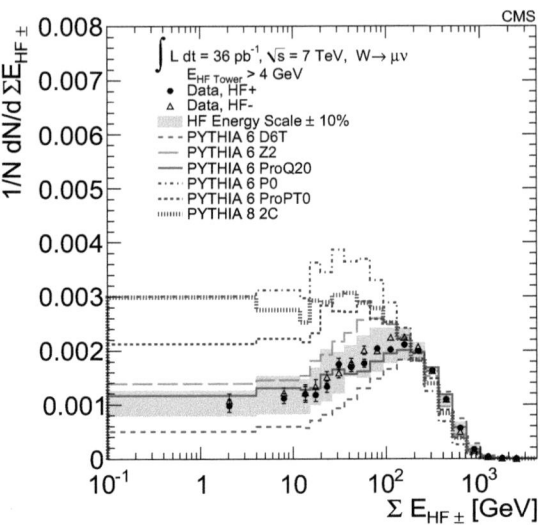

Figure 7.2: The summed HF+ and HF− energy distribution for $W \to \mu\nu X$ candidate events is shown for data and MC simulations with different tunes for the underlying event. The band indicates the uncertainty related to a ± 10% HF energy scale variation.

(about 0.05%) and the highest energy deposition is 2.7 TeV. All events with high-energy depositions also have large tower multiplicities. The average energies observed in HF+ and HF− are measured in 11 HF rings of calorimeter towers, each covering approximately a $\Delta|\eta|$ range of 0.175. The difference between the average energy deposition per ring (η bin) in the data and the different MC tunes is consistent for all tower rings. The average energy deposited per $|\eta|$ ring predicted by the D6T (Z2) tune is too large (too small), while the Pro-Q20 tune provides a very good description of the data.

The distributions of the total HF+ and HF− energy sums ($E_{HF+(-)}$) are shown in fig. 7.2 for the data and for several different MC tunes of the underlying event in $W \to \mu\nu$ events. The corresponding distributions to figs. 7.1a and 7.2 for $Z \to \mu\mu$ events are shown in fig. B.1.

The mean values of the reconstructed HF energy and the tower multiplicities for $W \to \mu\nu$ events are given in table 7.1, again for the three different instantaneous luminosity periods.

The average HF energy deposition, averaging the energy deposits in the HF+ and HF- calorimeters, in the data is 310 GeV, with an r.m.s. of 235 GeV. On average, about

Table 7.1: Mean energy depositions and tower multiplicities in each HF for single-vertex pp → W(→ $\mu\nu$)X events from the three running periods with different instantaneous luminosities.

Mean energy [GeV]	HF+ / HF−
P I	295.7 / 286.4
P II	313.0 / 295.7
P III	329.6 / 310.8
Average tower multiplicity	
P I	29.4 / 28.5
P II	30.9 / 29.6
P III	32.7 / 31.2

30 towers have an energy deposit of more than 4 GeV in each HF calorimeter. The statistical uncertainties of the mean energy values (mean tower multiplicities), estimated from the r.m.s. of the distribution, amount to less than ±5 GeV (±0.4) in the data and even smaller in the MC simulation. The corresponding mean value obtained with PYTHIA6 D6T, including the HF energy depositions from simulated pileup, is 370 GeV, with a tower multiplicity of 35. The PYTHIA6 Z2 tune predicts a mean energy deposition of 270 GeV and a tower multiplicity of 27, whereas using the Pro-Q20 tune results in a simulated energy deposition of 311 GeV and a tower multiplicity of 29 towers, similar to the data.

As can be seen from fig. 7.2, besides the Pro-Q20 tune, none of the MC models considered provides a good description of the HF energy distribution observed in the data. For energy depositions between 10 and 150 GeV, large differences between the data and different tunes are observed. In particular, the number of events in the data is about 30 to 50% higher than predicted by the D6T tune, and 50% lower than predicted by the Z2 tune. For simplicity, the older P0 and Pro-PT0 tunes are omitted from the following more detailed studies.

In total, 168 W and Z events in the muon decay channel, and 287 combined with the electrons, are found, where in one of the two HF calorimeters no individual tower reaches the energy threshold of 4 GeV, i.e., events with "zero" energy depositions in one HF. These events are defined as LRG events and are discussed in detail in chapter 8.

7.2 Soft Pileup Events and Forward Energy Distributions

The observed mean energy values in the HF increased by about 10 ± 5 GeV from period I to period II and by about 15 ± 5 GeV from period II to period III (cf. table 7.1). This increase of HF energy depositions is interpreted as arising from soft pileup events, not identified by the vertex finder. A similar increase of the mean energy deposition in HF is also seen in the MC simulations, when events with and without pileup are compared.

The properties of such soft pileup events have been studied with the zero-bias data samples, where the only requirement was that of beams crossing in the detector, taken during the different running periods. In events from this sample with zero reconstructed vertices, three classes of events can be identified: (1) events with no energy deposition in either HF (quasi-elastic pp-scattering), (2) events with zero energy in only one of the HF calorimeters and non-zero energy in the other (soft scattering with a LRG signature), and (3) events with non-zero energy depositions in both HF calorimeters (soft inelastic pp-scattering).

The contributions of beam-gas events and other beam-related backgrounds to the HF energy dispositions were studied in randomly triggered events with non-colliding beams and were found to be negligible.

The relative fraction of quasi-elastic pp-scattering event candidates in the zero-bias samples decreases from about 50% in period I to 20% in period III, while the fraction of soft inelastic events increases from about 15% in period I to 40% in period III. The fraction of soft events with a LRG signature of about 40% is roughly constant across the three luminosity periods. Hence, soft pileup events, not identified by the single-vertex requirement, can have an important effect on the HF energy distributions and on LRG events in particular. It was checked, that the contribution is well modeled by the MC simulations, by combining the contributions from the zero-bias to a non-pileup W sample.

7.3 Correlations of the Forward Energy Flow and the Central Charged-Particle Multiplicity

In the following, the correlation between the central charged-particle multiplicity and the forward energy flow is measured in the data and compared to MC models. For this study, events with energy depositions in the HF− calorimeter of 20-100 GeV (low), 200-400 GeV (medium), and above 500 GeV (high) are selected. The central charged-particle multiplicity distributions with track p_T thresholds of 0.5 GeV and 1 GeV, and the HF+ energy distributions for the three HF− energy intervals for W → $\mu\nu$ events are shown in fig. 7.3 and fig. 7.4, respectively. The corresponding distributions for Z → $\mu\mu$ events are shown in fig. B.2.

The charged-particle multiplicity distributions for the medium HF− energy range (fig. 7.3c) are described reasonably well by the PYTHIA6 D6T and Z2 tunes. The agreement between the data and the D6T tune is poorer when a 1 GeV track p_T threshold is applied. For the low HF energy range (fig. 7.3a), the D6T tune fails to describe the charged-particle multiplicity distribution, whereas the Z2 tune is in good agreement with the data, after applying a 0.5 GeV track p_T threshold. Finally, when requiring a large HF energy deposition (fig. 7.3e), the Z2 tune provides a good description, whereas the D6T tune overestimates the charged-particle multiplicity.

The HF+ distribution for the medium HF− energy interval (fig. 7.4b) is in better agreement with the predictions of the various tunes than the inclusive HF distribution (fig. 7.2). On the other hand, when requiring a low HF− energy deposition, the HF+ energy distribution is poorly modeled by all MC tunes (fig. 7.4a). Finally, for events with high-energy depositions in HF−, the PYTHIA8 generator, which in the inclusive case underestimates the rate of events with large HF− energy depositions, provides a good description of the energy distribution in HF+. Conversely, all other tunes predict more events with large HF+ energy than observed in the data (fig. 7.4c). To summarize these correlations, fig. 7.5 shows the HF+ versus HF− energy distribution (fig. 7.5a) and the HF+ energy versus the HF− track multiplicity distribution (fig. 7.5b).

Figure 7.6 shows the minimum and maximum energy deposition and tower multiplicity per event in the HF+ and HF− calorimeters for the data and the various MC tunes. In comparison to figs. 7.1 and 7.2, the differences between the data and all available MC tunes are somewhat enhanced. The corresponding figures for Z → $\mu\mu$ events are shown in fig. B.3.

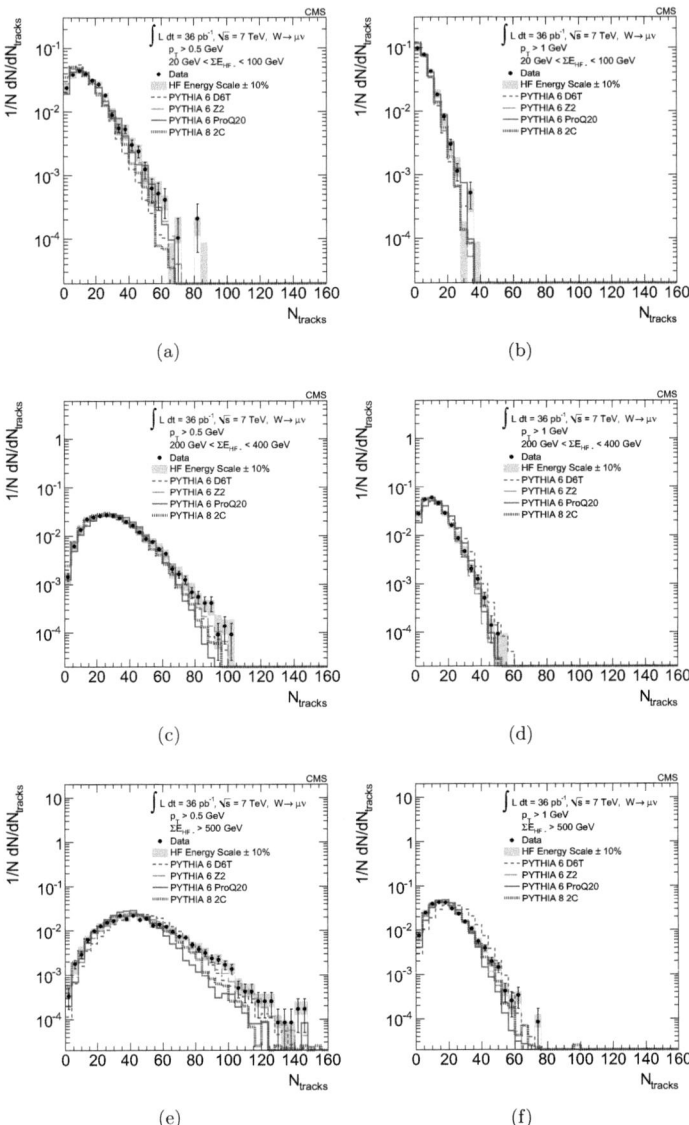

Figure 7.3: The charged-particle multiplicity distributions for two different p_T thresholds in the data and from MC simulations with different tunes, for the three HF− energy intervals of (a) and (b) 20-100 GeV, (c) and (d) 200-400 GeV, and (e) and (f) > 500 GeV.

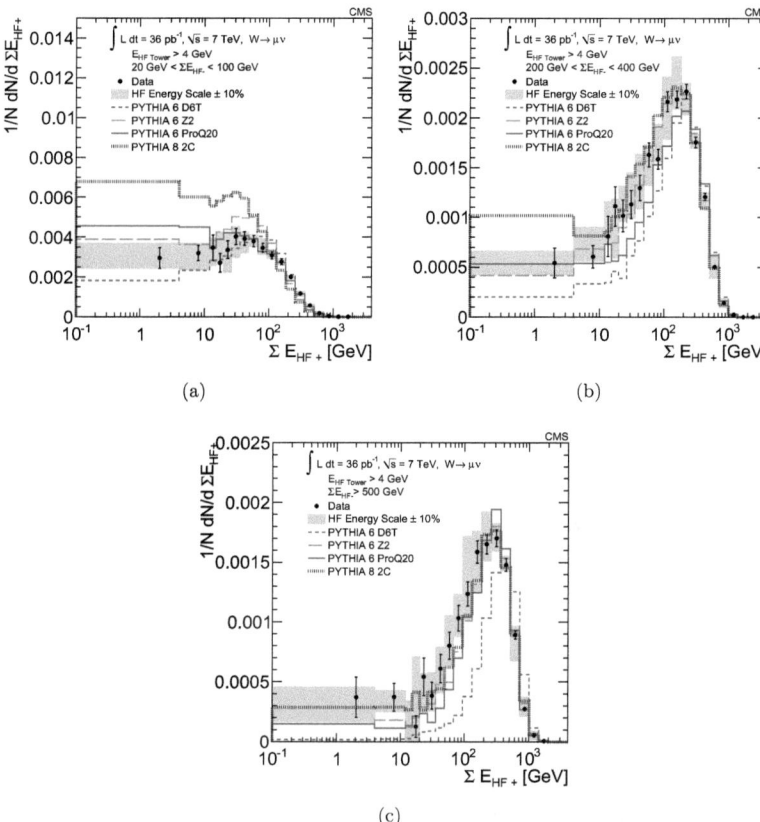

Figure 7.4: The summed HF+ energy distributions in the data and from MC simulations with different tunes, for the three HF− energy intervals of (a) 20-100 GeV, (b) 200-400 GeV, and (c) > 500 GeV.

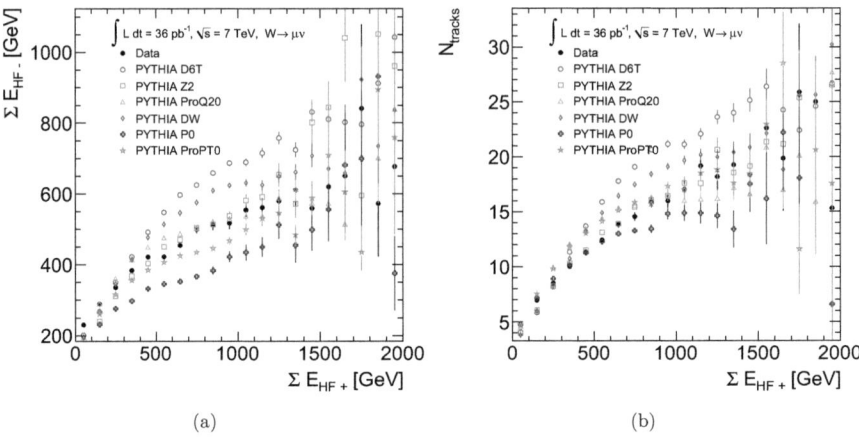

Figure 7.5: HF energy and charged-particle multiplicity distributions in $W \to \mu\nu X$ events for data and different MC tunes. (a) shows the correlation of the energy flow in HF+ and HF− and (b) the correlation of the energy flow in HF+ and the charged-particle multiplicity in HF−.

7.4 Interpretation of the Observed HF Energy and Charged-Particle Multiplicity Correlations

We have seen, that the energy distributions in the two HF calorimeters and the central charged-particle multiplicities are strongly correlated: large energy depositions in one of the HF calorimeters correspond to large energy depositions in the other HF calorimeter, as well as to an increase in the central charged-particle multiplicity. Such correlations are also predicted by the various MC tunes, though with very different strengths. The following conclusions can be drawn from the distributions:

D6T tune: The inclusive distribution of charged-particle multiplicities, with a minimum p_T of 0.5 GeV, is reasonably well described, whereas raising this threshold to 1 GeV leads to an overestimation of the event rate with large multiplicities. Furthermore, on average much-larger HF energy depositions are predicted than observed. When selecting events with small energy depositions in the HF, the fraction of events in the data is 30-50% larger than predicted by the D6T tune. In terms of correlations, the D6T tune provides a reasonable description only for the charged-particle multiplicity in the medium HF− energy interval (track p_T threshold of 0.5 GeV) and for the HF+ energy distribution

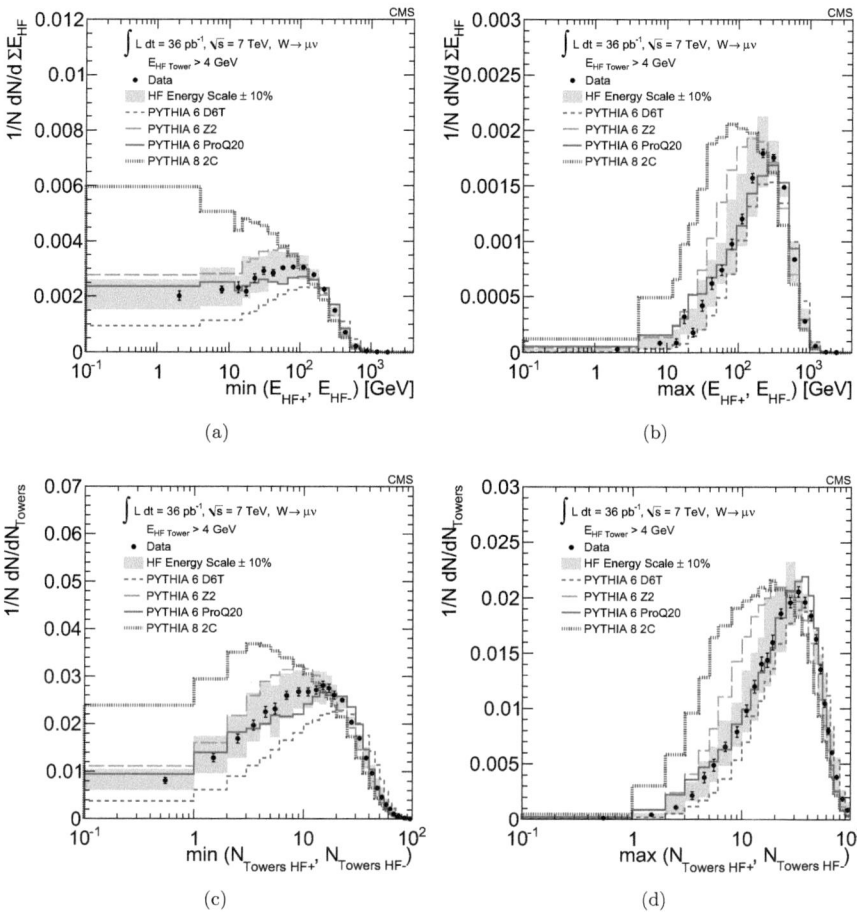

Figure 7.6: HF energy and tower multiplicity distributions in $W \to \mu\nu X$ events for data and different MC tunes. (a) and (c) show the minimum (min (E_{HF+}, E_{HF-}) and min ($N_{TowerHF+}$, $N_{TowerHF-}$, respectively) and (b) the maximum (max (E_{HF+}, E_{HF-} and max ($N_{TowerHF+}$, $N_{TowerHF-}$, respectively) of the energy depositions and the tower multiplicities per event in the HF+ and HF− calorimeters.

corresponding to the low HF− energy bin.

Z2 tune: Overall, the Z2 tune provides a very good description of the inclusive charged-particle multiplicities, but predicts too many events with very low charged-particle multiplicities. Concerning the HF energy distributions, too many events with low-energy depositions are predicted. The correlations between charged-particle multiplicity and HF− energy are well described. The HF+ energy distribution obtained for the low HF− energy interval is badly modeled, with the MC prediction much higher than the data at low energies. However, the correlations are well described for the higher HF− energy intervals.

Pro-Q20 tune: This tune provides the best description of the HF energy distributions and the charged-particle multiplicities with the $p_T > 0.5$ GeV threshold. However, the inclusive charged-particle multiplicity for the $p_T > 1.0$ GeV threshold is not well described, though still closer to the data than the D6T tune. In terms of correlations, the central charged-particle multiplicities are reasonably well described, though the fraction of events with large multiplicity and a large HF− energy deposition is underestimated. Furthermore, too many events with low-energy depositions in HF+ are predicted when a low-energy deposition in HF− is selected. For the other HF− energy bins this tune provides a good description of the data.

Pythia8 2C tune: In the inclusive case, this tune predicts too many events with low HF energy depositions, whereas the central charged-particle multiplicity distributions are well described. The HF+ energy distributions for the cases of low and medium HF− energy intervals are shifted towards lower values compared to data, whereas for the high-energy bin good agreement is found.

In summary, none of the analyzed MC tunes provides an overall consistent and reasonable description of the inclusive charged-particle multiplicities and the HF energy distributions in the W data sample, as well as correlations between them. It follows that the tunes, which provide a reasonable description of the underlying event structure for central rapidities in jet events, as presented in [57], require substantial modifications to describe the W data presented here. Similar, though statistically less significant results were obtained from the corresponding Z event samples.

7.5 Corrections to Hadron Level

The measurement of the forward energy flow can be very useful to improve current models of the underlying event in different MC tunes. Nevertheless, the distributions depend on the detector with which they have been measured. The limited resolution and efficiency of any detector induces a distortion on the observed variable characteristic of the detector. This detector dependance has to be taken into account before any comparison to other experiments or theories can be done. Several solutions to this problem exist. The simplest and quickest approach is to determine a response or smearing matrix A which can be included in a specific theory and provides a direct comparison to the data, i.e. the comparison of the theory to the measured data is done on the detector level

$$Ax = y, \qquad (7.1)$$

where x corresponds to the true value and y is the measured data. If A is known, Ax can be determined for a certain theory and be compared to the measurement y.

A second, more tedious approach is the inverse process (unfolding) where the unfolding matrix A^{-1} has to be determined. The unfolding matrix corrects for the distorted data at the detector level and provides a direct comparison to any theoretical prediction or corrected experimental data. However, the determination of such a matrix can be extremely difficult; the response matrix has to be inverted which introduces numerical instabilities and a regularization procedure has to be implemented. Without a proper treatment, small changes in the measured distribution can result in very large errors on the corrected distribution which is compared to the theory.

Nevertheless, for most of the measured data which are compared to a theory it suffices to determine the smearing matrix A and run the MC generator to convert hadron level to detector level. This approach is chosen for the measurement of the forward energy flow with centrally produced W or Z bosons.

7.5.1 Cuts on Generator Level

With a perfect detector, the entries of the response matrix would correspond to one along the diagonal and zero for the rest of the entries. This will not be the case for a real detector, but the correction factors should be small enough, i.e. close to one and zero, respectively. Figure 7.7 shows the summed HF energy distribution for simulated events generated with the PYTHIA6.4 MC event generator for two different tunes, D6T and Z2, and for the events passed through the CMS detector simulation based on GEANT4. The generator level distribution is at the stable-particle ($\tau > 10^{-12}$ s) level. The same

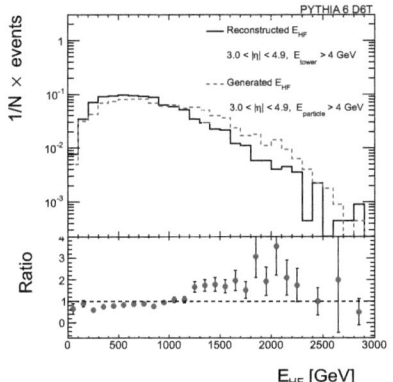 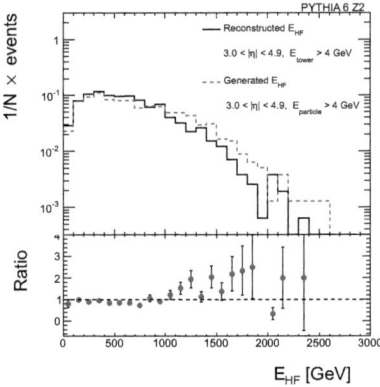

Figure 7.7: The summed HF energy distribution for $W \to \mu\nu$ events at generator and detector level. The left plot shows the prediction from the PYTHIA6 D6T event generator tune and the right plot the prediction from the PYTHIA6 Z2 tune. In the lower panel the ratio of generated events over reconstructed events is shown.

acceptance cuts are applied on generator level as for the data analysis chain:

- muon transverse momentum $p_T > 25$ GeV

- muon pseudorapidity $|\eta| < 1.44$

- neutrino transverse momentum $p_T > 30$ GeV

- W transverse mass $m_T > 60$ GeV

Additionally to the acceptance cuts, one vertex is required and the z-position of the vertex has to be within 2 cm of the detector point of origin. The energy cutoff of 4 GeV is applied at both levels; at detector level it is applied on the single towers and at generator level it is applied to the single particles. The η acceptance is from $|\eta| = 3.0$ to $|\eta| = 4.9$. The ratio between generator and detector level distribution varies between 1 and 2 for an energy up to 1.5 TeV, where the statistical error is still relatively small (smaller than ±0.4). The ratio of the mean is 1.2 for the D6T tune and 1.1 for the Z2 tune.

7.5.2 Bin-by-bin Correction Factors

To account for the geometry of the HF detector with it's division in rings (see chapter 3), the correction to hadron level is done in five η intervals. The intervals are chosen such that each one contains two rings. Table 7.2 shows the pseudorapidity range corresponding to the five η intervals.

Table 7.2: Pseudorapidity ranges for the five bins corresponding to the rings of the HF detector used in the summed HF energy distributions.

	η_1	η_2	η_3	η_4	η_5
Range	3.1 - 3.5	3.5 - 3.8	3.8 - 4.2	4.2 - 4.5	4.5 - 4.9

Figure 7.8 shows the distributions for the five pseudorapidity ranges. The distributions are normalized to one.

The inner rings, corresponding to η_2 to η_4 show quite a good agreement, whereas in the outer rings (η_1 and η_5) the distributions at generator level have significantly larger tails compared to the detector level. Several detector-related effects can be responsible for this difference: energy resolution and non-linear response in the HF detector, migration effects between different pseudorapidity bins due to scattering on dead material or deviation of the particle-track out or in to the adjacent η-bin due to the magnetic field. The energy resolution and non-linear response has been checked with test beam data and found to be properly simulated. On the other hand it was found that the largest effect on migration comes from varying the material budget, resulting in differences particularly in the region closest to the beam pipe, i.e. for the bin η_5. We can roughly correct for this effect with a constraint in the acceptance on the generator level with respect to the reconstruction level. Figure 7.9 shows the same distribution as fig. 7.8 for the η ranges 1 and 5 including the distribution with a corrected acceptance on generator level of $3.2 < |\eta| < 3.5$ and $4.5 < |\eta| < 4.8$, respectively.

The agreement between the reconstructed and generated distributions improves significantly with the modified pseudorapidity acceptance, especially for the outer bin η_5 (lower plots in fig. 7.9). Evidently the same acceptance corrections should be applied to the other bins, but from the already good agreement of the distributions one can assume that the bin-to-bin migrations between the η-bins are relatively small.

Other influences on the reconstructed shape have been studied but none of them resulted in a significant distortion of the shape. The following list summarizes these studies.

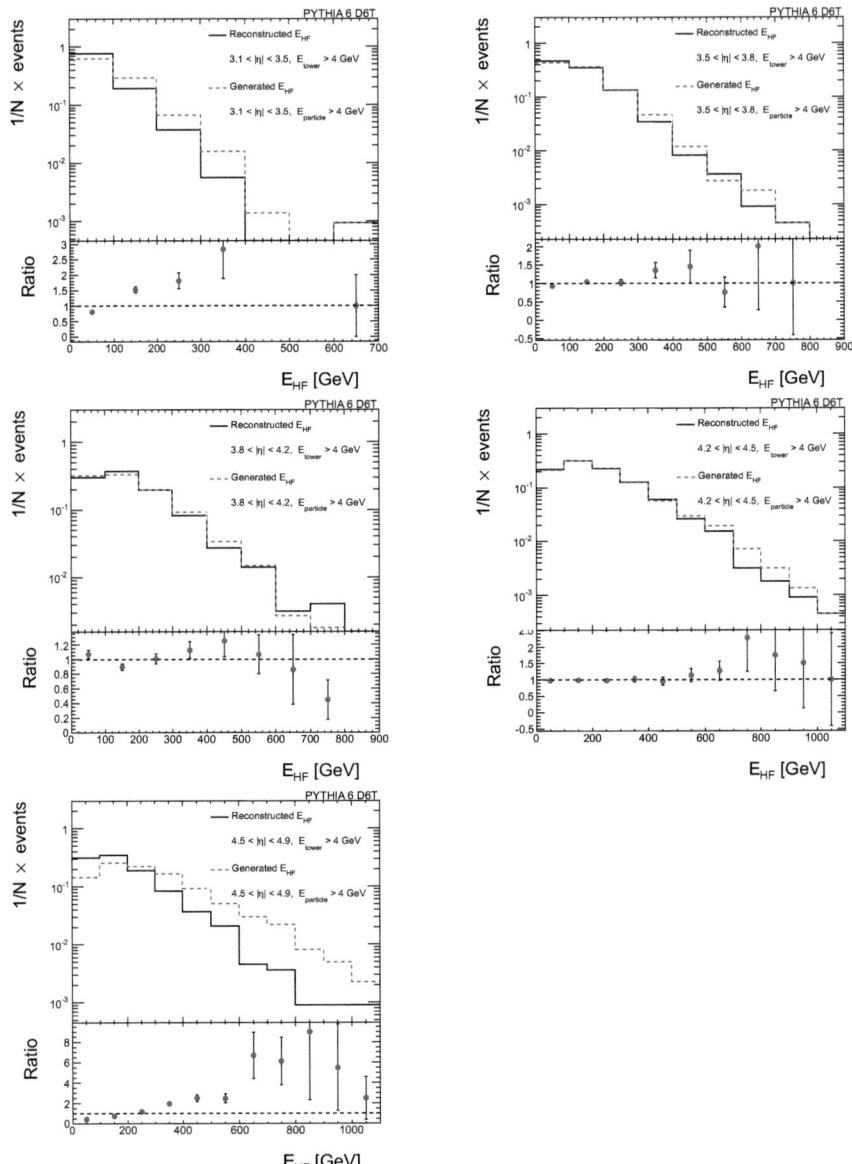

Figure 7.8: The summed HF energy distribution for $W \to \mu\nu$ events at generator and detector level for the PYTHIA6 D6T event generator tune and for the five pseudorapidity ranges η_1 - η_5. In the lower panel the ratio of generated events over reconstructed events is shown.

Forward Energy Flow and Central Charged-Particle Multiplicity

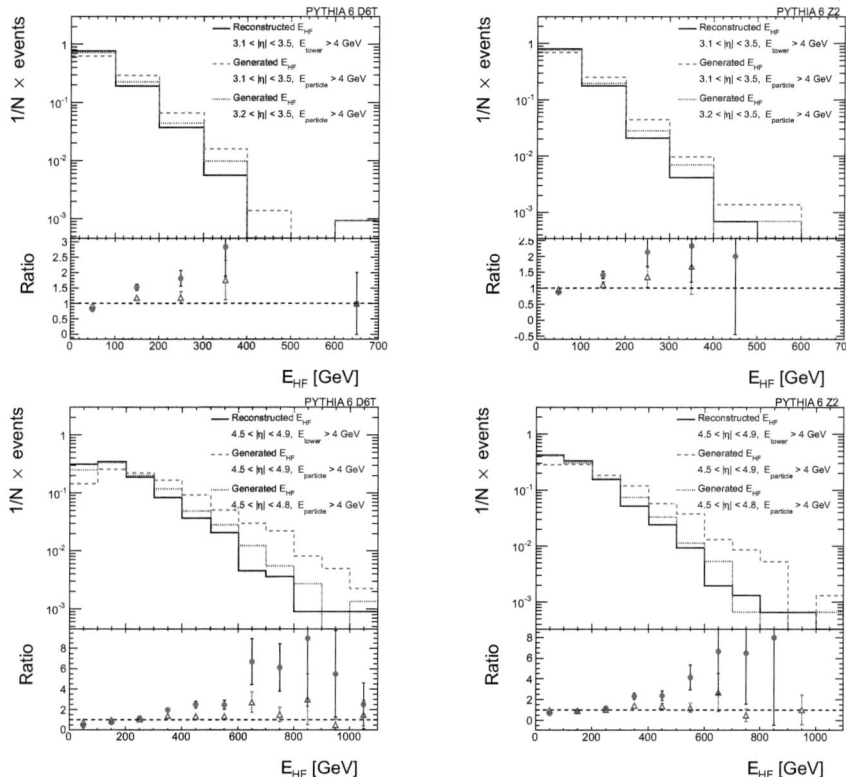

Figure 7.9: The summed HF energy distribution for $W \to \mu\nu$ events at generator and detector level for the PYTHIA6 D6T event generator tune on the right and for the PYTHIA6 Z2 tune on the right for the constrained pseudorapidity acceptances for η_1 (upper plots) and η_5 (lower plots). In the lower panel the ratio of generated events over reconstructed events is shown.

- **HCAL vs. ECAL energy reconstruction:** The shower size and form of the particles reconstructed in the ECAL (electrons and photons) and in the HCAL (hadrons) differ from each other: the shower shape in ECAL is much shorter and broader than in the HCAL. For this reason the resolution of the energy reconstruction in the former is better than in the latter and the larger spread of the showers in HCAL could lead to a distortion of the shape. Applying a cut on energy only on the hadronic particles leads as expected to an overall shifted energy distribution to the lower values, but does not remarkably improve the shape.

- **Charged and neutral particles:** Another possible flaw could be in wrongly reconstructed charged and/or neutral particles. It was checked, that the proportions of charged and neutral particles are the same for the events where the agreement was good and for events with a difference in energy larger than 100%, $\Delta E = E_{gen} - E_{reco} > 100\%$. No deviation from the expected ratio could be observed.

- **Large energies of single particles:** Single particles with very large energy might also lead to a distortion in the shape. This can very easily be checked by applying a cut on all particles at generator level with very large energy. No evidence of shape distortion caused by particles with an energy above 500 GeV was found.

Figure 7.10 shows the energy flow as a function of η for the five η-bins for the PYTHIA6 event generator tunes D6T and Z2 for the corrected and the uncorrected acceptance on the generator distribution.

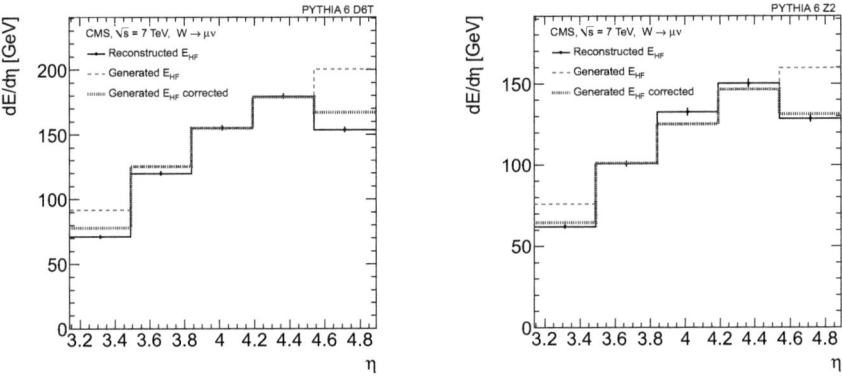

Figure 7.10: Energy flow as a function of η in $W \to \mu\nu$ events for the PYTHIA6 D6T (left) and Z2 (right) event generator tunes. The Black histogram corresponds to the reconstructed energy and the colored lines to the generated events. The blue histogram is acceptance corrected in the inner- and outermost bins.

Finally, the correction factors are calculated bin-by-bin as the ratio of the distributions at the generator level over the detector level, for both tunes and before and after the modification of the acceptance on generator level (fig. 7.11). The factors range from 0.9 to 1 for the D6T tune and between 0.96 and 1.06 for the Z2 tune. The Z2 factors are slightly higher than the D6T, and both show the same small difference in shape for the detector and the stable-particle level distribution.

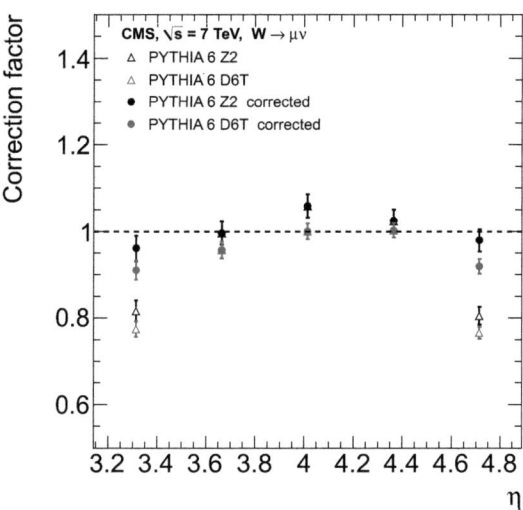

Figure 7.11: The correction factors for the five η-bins, calculated from the ratio of MC HF energy predictions at the stable-particle level and the detector level. Black dots and triangles correspond to the PYTHIA6 Z2 tune, red dots and triangles to the PYTHIA6 D6T tune. Dots correspond to the energy with acceptance correction in the inner- and outermost bins, triangles to the uncorrected ones.

7.5.3 Response Matrix

Additionally to the bin-by-bin correction factors from the previous section the n×n smearing matrix can be determined. The same cuts are applied on generator level as in the previous section. The result is shown in fig. 7.12. The entries of the matrix are normalized to one such that the sum of all the bins for the reconstructed energy equals to the value of the corresponding generated energy, in percent. The largest part of the energy is close to the diagonal and only small amounts are on the far sides of it. There is a tendency towards smaller reconstructed energies in agreement with the results obtained in the previous section. Above 800 GeV the entries are zero for the Z2 tune due to the lack of statistics.

For a better visualization of the distribution of the matrix entries a two-dimensional colored plot is displayed in fig. 7.13 and a three-dimensional colored plot in fig. 7.14. The conclusions mentioned before can be observed very clearly in these plots.

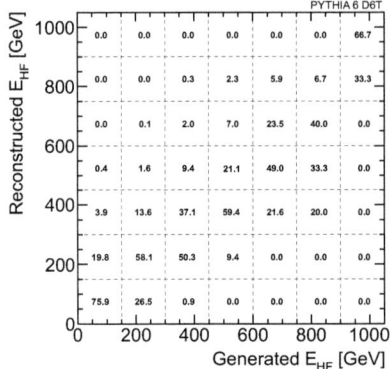
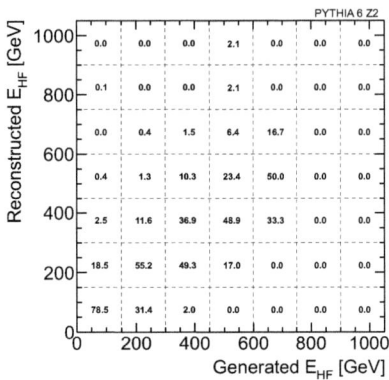

Figure 7.12: The response matrix for the forward energy flow in events with W bosons for the PYTHIA6 D6T event generator tune on the left and for the PYTHIA6 Z2 tune on the right. The entries in the matrix are in percent and normalized such that the sum of all entries on the y-Axis corresponding to a certain x-bin is 100%. Above 800 GeV the entries are zero for the Z2 tune due to the lack of statistics.

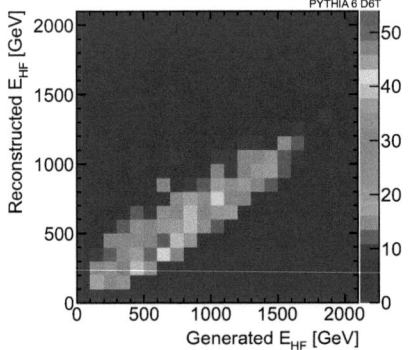
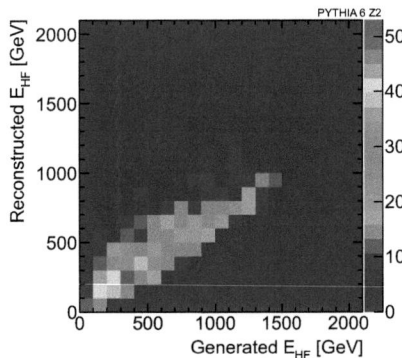

Figure 7.13: Visualization of the response matrix in a two-dimensional colored plot for the PYTHIA6 D6T event generator tune on the left and for the PYTHIA6 Z2 tune on the right.

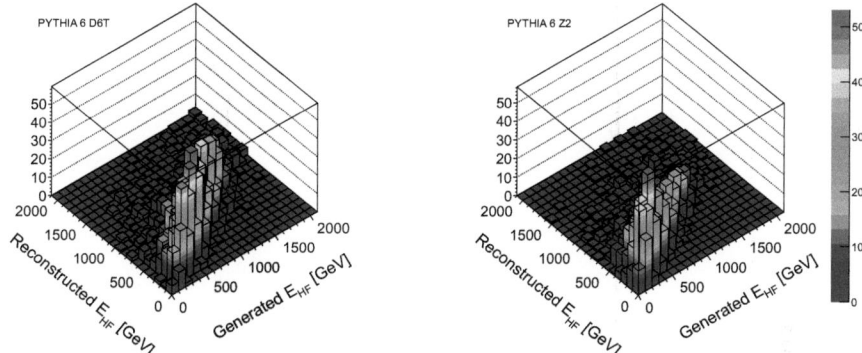

Figure 7.14: Visualization of the response matrix in a colored lego plot for the PYTHIA6 D6T event generator tune on the left and for the PYTHIA6 Z2 tune on the right.

Chapter 8

W and Z Events with a Large Pseudorapidity Gap

A subset of W and Z events with a single primary vertex and a LRG signature is studied in this chapter. A LRG event is defined by the requirement that none of the calorimeter towers has an energy deposit of more than 4 GeV in at least one of the two HF calorimeters, corresponding to a pseudorapidity gap of at least 1.9 units. Several distributions and characteristics of this subset are discussed, with the final aim to introduce a key variable for a clear distinction of non-diffractive events from events with a diffractive component in them, in the last section. The results in this chapter are presented for a combined measurement of muons and electrons in the final state, which almost doubles the number of events.

8.1 Observed Number of LRG Events

Table 8.1 shows the observed LRG event yields and their ratio to the number of inclusive W and Z single-vertex events for the three luminosity periods. This ratio decreases by roughly a factor of 2 to 4 when going from period I to period III. The decrease can be explained by the HF energy depositions coming from soft pileup events. As discussed before, pileup in the Monte Carlo simulation shifts some of the LRG events to the class of low-energy depositions in the HF. The fraction of events with no detected vertex but beam on both sides and energy deposits in the HF detectors has been determined from the zero bias data. This sample includes beam related background like beam gas interaction which can also increase with increasing instantaneous luminosity. The number of LRG

Table 8.1: Number of LRG events with a single vertex and their percentage relative to all selected W and Z events, for the three different luminosity periods and their total.

	$W \to e\nu$	$W \to \mu\nu$	$Z \to ee$	$Z \to \mu\mu$
Total	100 (0.71%)	145 (0.81%)	19 (0.80%)	23 (0.79%)
P I	17 (1.13%)	31 (1.61%)	7 (2.70%)	3 (0.91%)
P II	57 (0.72%)	91 (0.86%)	9 (0.59%)	16 (0.93%)
P III	26 (0.57%)	23 (0.42%)	3 (0.55%)	4 (0.46%)

events are corrected by this fraction according to

$$f_{\text{corr}} = \frac{1}{1 - (0.5 \cdot p_1 + p_2 + p_{\text{too close}})}$$

where p_1 is the probability for a soft pileup with undetected vertex with energy deposits in one of the HF detectors, p_2 as p_1 but energy in both HF detectors and $p_{\text{too close}}$ the probability for a soft pileup with its vertex too close to the W vertex to be separated and detected. If an undetected pileup event leaves energy deposits in both HF detectors (p_2) then it is impossible to detect the LRG event. On the other hand if a pileup event leaves energy in only one of the HF detectors (p_1) in 50 % of the LRG events the gap can be detected, hence the factor 0.5 in front of p_1. The probability factor $p_{\text{too close}}$ is determined with the obtained inefficiency to detect the merged pileup vertex $\epsilon = 3.3\%$ (cf. section 6.2):

$$p_{\text{too close}} = \epsilon \cdot P(1, \lambda) + \epsilon \cdot P(2, \lambda) \cdot (1 - \epsilon) + \epsilon \cdot P(3, \lambda) \cdot (1 - \epsilon)^2$$

where the higher order terms of ϵ are negligible. $P(i, \lambda)$, i = 1, 2, 3 refers to the Poisson probability function with the average number of pileup events λ expected in the periods P I to P III.

After correcting the observed number of LRG events in the data for pileup effects a constant fraction of LRG events, relative to the total number of W and Z events with a single primary vertex, is found for the three instantaneous luminosity periods. The corrected fraction of LRG events is given in table 8.2 for muons and electrons. The uncertainties on this correction are small compared to the statistical errors and to the ±10% energy scale uncertainties of the HF calorimeters (for details see [122]). This energy scale variation is the dominant systematic uncertainty for the estimated fraction of LRG events in the data, resulting in a change of about ±26% when varying the tower energy threshold between 3.6 and 4.4 GeV.

Table 8.2: Percentage of LRG events in single-vertex W and Z events for muons and electrons, using a pileup correction determined from data, for the entire dataset and the three different luminosity periods. Only the statistical uncertainties are given; the dominant systematic uncertainty from the HF energy scale is about ±26%.

	W → eν	W → μν	Z → ee	Z → μμ
Total	1.37 ± 0.14%	1.50 ± 0.13%	1.73 ± 0.43%	1.49 ± 0.31%
P I	1.68 ± 0.40%	2.39 ± 0.43%	5.52± 2.08%	1.36 ± 0.78%
P II	1.27 ± 0.17%	1.54 ± 0.16%	1.57 ± 0.52%	1.65 ± 0.41%
P III	1.53 ± 0.30%	1.12 ± 0.23%	1.47± 0.85%	1.22 ± 0.61%

Combining the results obtained with electrons and muons, the percentage of W and Z events with LRG signature is (1.46 ± 0.09 (stat.) ± 0.38 (syst.))% and (1.57 ± 0.25 (stat.) ± 0.42 (syst.))%, respectively. In comparison, as can be seen from figs. 7.1, 7.2 and 7.6, the fraction of W LRG events predicted with the PYTHIA6 Z2 and Pro-Q20 and the PYTHIA8 2C tunes are larger than observed in the data. In contrast, for the D6T tune the number of LRG events is smaller than in the data.

8.2 Jet Activity in W/Z Events with a LRG Signature and Search for Exclusive W/Z Production

A further subset of W and Z events are those that show some jet activity, using the particle-flow algorithm with a cone size of 0.5. We find that $(11.1 \pm 0.2)\%$ of the selected W and Z events with a single vertex contain at least one reconstructed jet with a transverse momentum above 30 GeV and $|\eta| < 2.5$. Taking the subsample of 145 identified W → μν events with a LRG signature, 8 events are found with one or more jets above a 30 GeV threshold. The corresponding numbers for the electron channel are 100 identified LRG events and 8 of them with at least one jet. Thus, we find that (6.5±1.6)% of the LRG events have jet activity, which is smaller (but still consistent) with the fraction of events with jets observed in the inclusive W sample. No other particular features of the events with jet activities, when compared to the MC simulations, are observed.

Another potentially interesting class of LRG events consists of W and Z events with essentially no activity besides that from the vector-boson decays. Such candidate exclusive events are selected with the requirement that both HF calorimeters fulfill the LRG condition and that no particle-flow object besides the lepton(s) is reconstructed in the central detector, above a transverse momentum threshold of 0.5 GeV. For the electron selection no such events are found. In the muon case, 2 W and 2 Z event candidates with zero energy in both HF calorimeters are found. In fig. A.3 one of the Z candidate events can be seen in an event display of the detector. All four events have some reconstructed tracks in the central detector with small energy. The number of observed events is consistent with the expected number of non-exclusive W and Z events predicted from the MC simulations.

8.3   Size of the Pseudorapidity Gap and Central Gaps

For the study of events with large pseudorapidity gaps an interesting parameter is how far the size of the gap extends into the central detector. One might intuitively expect that the ratio of diffractive signal events compared to background from multiplicity fluctuations, would become larger when the gap size increases into the central detector. This intuitive view is confirmed when comparing diffractive W events simulated with POMPYT, where the decrease in event yields with increasing gap size is much smaller than in the different non-diffractive MC models.

The definition of the gap is ambiguous, as the meaning of zero activity or zero energy depositions depends on the experimental criteria for the detection of particles in the data and in the MC simulation. For this study the size of the pseudorapidity gap was determined by using particle-flow objects with a minimum energy of 1.5 GeV for $|\eta| < 1.5$ (barrel), 2 GeV for $1.5 < |\eta| < 2.85$ (endcaps), and 4 GeV for $|\eta| > 2.85$ (HF calorimeters). For charged particle-flow candidates a minimum transverse momentum of 0.5 GeV was required. The largest (η_{max}) and smallest (η_{min}) observed pseudorapidity values of the particles are used to determine the gap size between the maximum (minimum) η coverage of the experiment and the nearest detected particle on each side. In order to combine both hemispheres, we define $\tilde{\eta}$ as the minimum of η_{max} and $-\eta_{min}$. The size of the pseudorapidity gap is then $\Delta\eta_{gap}^{4.9} = 4.9 - \tilde{\eta}$, where 4.9 is the largest η value covered by the HF.

Fig. 8.1a shows the $\tilde{\eta}$ distribution in the data and MC simulation with different tunes for the W $\to \mu\nu$ events. The fraction of events with pseudorapidity gaps decreases rapidly when the gap size increases. A statistically significant excess of events at $\tilde{\eta}$ values smaller than 3 is observed, compared to the predictions of the non-diffractive model implemented in PYTHIA6, tune D6T. However, the fraction of events with very large gaps and without a diffractive component in the PYTHIA6 Z2, Pro-Q20, and PYTHIA8 2C tunes is larger than in the data.

The stability of the $\tilde{\eta}$ distribution was tested by allowing a ±10% variation of the particle-flow candidate energy and momentum thresholds in the data. The resulting variations were found to be similar to the statistical uncertainties.

If $\tilde{\eta} < 0$, all the reconstructed particle-flow objects in the event are contained in one hemisphere. Combining the W events with LRG signature in both lepton channels, 4 events with one "empty" detector hemisphere, corresponding to a gap of at least $\Delta\eta_{gap}^{4.9} = 4.9$ units in pseudorapidity, are observed. In comparison, 0.8, 3.5, and 2.2 such events are expected from the non-diffractive MC simulation based on the PYTHIA6 D6T, Z2, and Pro-Q20 tunes, respectively.

As discussed before, soft pileup events without a detectable second vertex remain in the sample. However, since these events do not produce significant particle-flow in the central pseudorapidity regions, the effect on events with pseudorapidity gaps in the more central region, $|\eta| < 2.85$, is expected to be small. The number of pseudorapidity gap events in this detector region, when ignoring the information in the HF detectors, should mainly depend on the amount of very low multi-parton activity and thus on the number of low-multiplicity events. The $\tilde{\eta}$ distributions using only particle-flow objects with $|\eta| < 2.85$ and ignoring the information in the HF are shown in fig. 8.1b. Accordingly, the gap size is now defined as $\Delta\eta_{gap}^{2.85} = 2.85 - \tilde{\eta}$. Again, when compared to the MC simulation with the D6T tune, the data show a large excess of events with $\tilde{\eta}$ below 1, corresponding to a central pseudorapidity gap of $\Delta\eta_{gap}^{2.85} \geq 1.85$. The fraction of such gap events in the data is reasonably well described by the PYTHIA6 Z2 and Pro-Q20 tunes, and much larger fractions are predicted by the PYTHIA8 2C tune.

The limited number of LRG events, as well as the large uncertainties related to the modeling of the underlying event and multi-parton interactions, prevent any conclusions from being drawn on the possible presence of a diffractive W/Z-production component from the observed rate of events with a pseudorapidity gap in the central detector.

8.4 Charged-Particle Multiplicity and Forward Energy Distributions in LRG Events

The charged-particle multiplicity distribution in LRG events, from combining the W events in the electron and muon channels, is shown in fig. 8.2a for a minimum track p_T of 0.5 GeV. A slight excess of events with large charged-particle multiplicities is found in the data, compared to the various non diffractive MC tunes. However, overall the

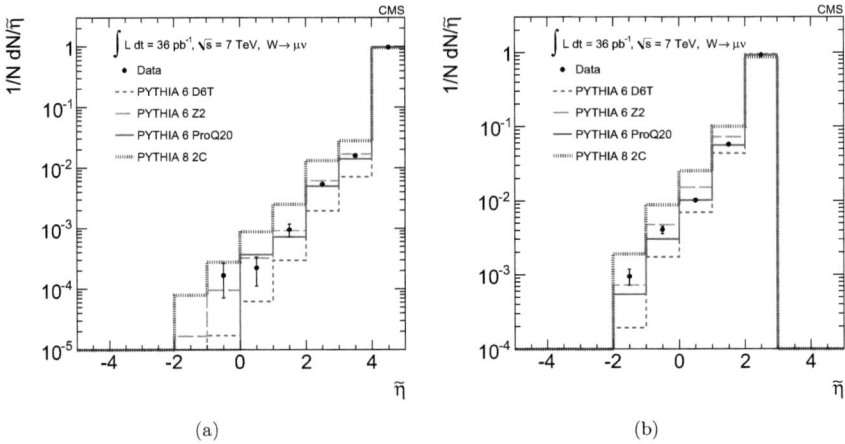

Figure 8.1: (a) The $\tilde{\eta}$ distribution for W events with muon decays in data and for various MC simulations and (b) the corresponding distributions ignoring the HF calorimeter information.

track p_T spectrum is well described. The number of LRG events with 20 and more tracks, combining the electron and muon channels, is 33 in the data. Only 13 (19) events with more than 20 tracks are expected from the D6T (Z2) tunes. A similar, but statistically less significant, excess of events with multiplicities larger than predicted by the different tunes is also observed when a track threshold of $p_T > 1.0$ GeV is required.

The POMPYT diffractive model, which does not include multi-parton interactions, predicts even smaller charged-particle multiplicities. However, the observed excess of events with relatively large charged-particle multiplicities in LRG events could be an indication of a diffractive component in the multi-parton interactions, as depicted in figs. 6.2c and 6.2f.

The corresponding distribution for the energy sum in the HF calorimeter opposite to the gap is shown in fig. 8.2b. The average total energy of 150 GeV with an r.m.s. of 160 GeV is about a factor of two smaller than the one observed for the inclusive HF energy distribution, and is reasonably well described by the various MC tunes. In the data, we find 2 events with no towers above the energy threshold of 4 GeV in either HF calorimeter, in agreement with the expectation from the D6T tune. This number is slightly lower, but still consistent, with the expections from the Z2 and the PYTHIA8 2C tunes. The Pro-Q20 tune predicts 8 such events.

W and Z Events with a Large Pseudorapidity Gap 119

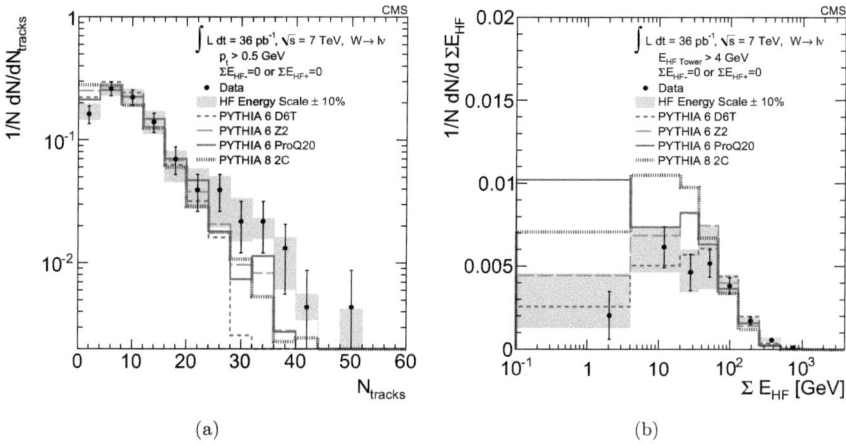

Figure 8.2: (a) Charged-particle multiplicity and (b) HF energy distributions (opposite to the gap) in pp $\to$ W$^\pm X \to \ell^\pm \nu X$ events with a LRG signature, for the data and different MC tunes. The charged-particle multiplicity distribution is obtained for a track p_T threshold of 0.5 GeV.

8.5 Hemisphere Correlations Between the Gap and the W (Z) Boson

Figure 8.4 shows the distribution of the signed charged lepton pseudorapidity η_ℓ in W events with a LRG signature (electron and muon channels combined). The sign is defined to be positive when the gap and the lepton are in the same hemisphere and negative otherwise. The data show that charged leptons from W decays are found more often in the hemisphere opposite to the gap. Combining the electron and muon channels, 147 events are found with the charged lepton in the hemisphere opposite to the gap and 96 events with the lepton in the same hemisphere. Defining an asymmetry as the ratio of the difference between the numbers of LRG events in each hemisphere and the sum, the corresponding asymmetry is (-21.0 ± 6.4)%. In the case of Z candidates (the rapidity of the Z is used) with a LRG signature, 24 (16) events are in the opposite (same) hemisphere as the gap, resulting in an asymmetry of (-20 ± 16)%.

In comparison, the various non-diffractive MC tunes predict a symmetric lepton pseudorapidity distribution in LRG events. On the other hand, events generated with the POMPYT generator, based on a diffractive production model, exhibit a strong asymmetry.

This can be explained in terms of diffractive PDFs, which peak at smaller x than the conventional proton PDFs (fig. 8.3). The produced W(Z) is thus boosted in the direction of the parton that had the larger x. This is typically the direction of the dissociated proton, i.e., opposite to the gap.

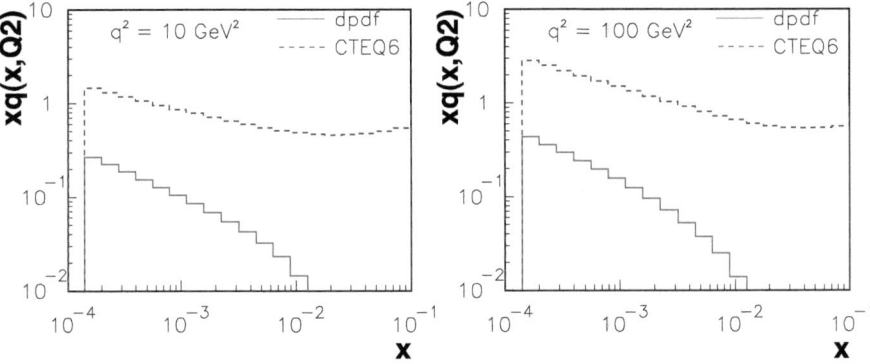

Figure 8.3: Diffractive parton distribution function (dpdf) and conventional proton PDF (CTEQ6) for two different q^2 values illustrating the different behavior for large x values leading to a boost of the W (Z) in the direction of the dissociated proton [123].

The signed lepton pseudorapidity distribution in the data is fit to the predictions from the diffractive POMPYT and the non-diffractive PYTHIA event generators, with the relative fraction of the two as the free parameter. The fit results in a fraction of diffractive events in the LRG sample of $(50.0 \pm 9.3$ (stat.) ± 5.2 (syst.))%, assuming the model of diffraction implemented in POMPYT and using the PYTHIA6 Pro-Q20 for the simulation of non-diffractive events. The fit results are shown in fig. 8.4. The fits using the combination of POMPYT with other tunes give similar results, and only the non-diffractive contribution from the other tunes is shown in fig. 8.4. The systematic uncertainty of 5.2% has been determined from the 10% HF energy scale variations and from the fits with the different tunes, using the maximal and minimal fractions obtained from the different fits.

The asymmetry in the signed η_ℓ distribution for non-LRG events is found to decrease when the forward energy deposition increases. For example, for HF energy depositions in the intervals 20-100 GeV, 200-400 GeV, and > 500 GeV, the asymmetry is $(-3.5 \pm 1.1)\%$, $(-2.7 \pm 1.0)\%$, and $(0.9 \pm 2.3)\%$. The small residual asymmetry in events with low HF energy depositions is insignificant in comparison to the one in LRG events. However, this could be explained by the presence of a small fraction of diffractively produced W boson events in which the LRG signature is destroyed by the accompanying multi-parton interaction or by the undetected pileup component. For higher energy depositions in the forward region, the asymmetry vanishes.

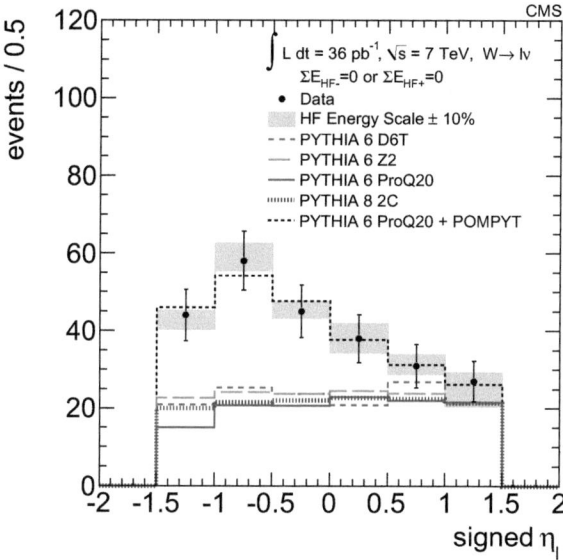

Figure 8.4: Signed lepton pseudorapidity distribution in W events with a LRG signature, with the sign defined by the pseudorapidity of the lepton relative to the gap (positive for the lepton and gap in the same hemisphere, negative otherwise). Electron and muon channels are combined. The fit result for the combination of PYTHIA6 (Pro-Q20 tune) and POMPYT predictions is shown as a dotted black line. For the other PYTHIA tunes, only the non-diffractive component is shown.

Chapter 9

W and Z Transverse Momentum Spectrum

The W/Z transverse momenta in proton collisions arise from the initial state radiation of the interacting partons. Thus, the bosons p_T distributions are sensitive to the dynamics of the parton-parton collision. The approaches to describe the low and high momentum are different and therefore the study of each of these regions provides a different QCD test of the used models and calculations.

In the low p_T region, roughly below 10 GeV, the process is dominated by soft gluon radiation and fixed order perturbative calculations can not be applied. Resummation of large logarithms is necessary in combination with non-perturbative descriptions. The current approach to describe non-perturbative processes is based on the tuning of parameters in different MC models to data. Thus, the low p_T region is sensitive to underlying processes and precision measurements can provide a better understanding.

In the high p_T region, roughly above 30 GeV, the process is dominated by the radiation of a hard parton and perturbative calculations can be applied. The production cross sections are obtained using as input the PDFs from global fits to data. The total and differential W/Z cross sections have been calculated to NNLO [124, 125] and the dominant errors remain in the uncertainties on the PDFs, the strong coupling constant α_s and the renormalization and factorization scales. For example, the uncertainties on the production cross sections of the W, Z at the LHC due to the PDFs are estimated to be ±3.5%[126]. Therein, the gluon distribution function has the greatest uncertainty. This can be seen in fig. 9.1, where the q$\bar{\text{q}}$ and gluon-gluon luminosity uncertainties from the CTEQ6 PDF generation are shown.

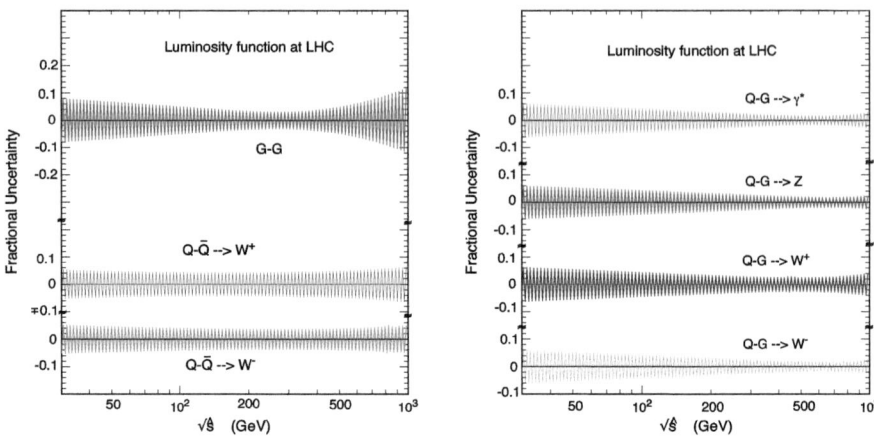

Figure 9.1: Fractional uncertainties of the parton-parton luminosity functions at the LHC[19]

Therefore, the precise measurement of the PDFs is an important input for accurate cross section predictions and validations of the standard model. The measurement of the high p_T region is sensitive to the gluon distribution functions. To illustrate this, a qualitative description of the PDF influence on the p_T spectrum is depicted along with some exemplary figures. Afterwards, the results of the W and Z p_T measurements are presented for high and low p_T.

9.1 Transverse Momentum and PDFs

The p_T spectra obtained with two widely used PDF sets (MSTW2008 and CTEQ6L1) are compared in WX $\rightarrow \mu\nu$X and ZX $\rightarrow \mu\mu$X events. For conciseness, the distributions are shown either for W or for Z events; the conclusions are the same for both event classes.

Figure 9.2 shows a comparison of the Z transverse momentum for the CTEQ6L1 and MSTW2008 PDF sets. We can see that the ratio of the two PDFs differs significantly from one. Above 10 GeV the CTEQ based spectrum has roughly 8% more events than with MSTW. The ratio is constant above 30 GeV, within the errors, with an excess of CTEQ over MSTW of around 10%. The event yield dependance on p_T points out the sensitivity to the composition of the production processes.

The jet in the leading order picture for W/Z+jet production can originate from a gluon

("gluon-induced") or a quark ("quark-induced") (section 2.2.3). The composition of these production processes varies as a function of p_T but also for different PDFs. This can be seen in fig. 9.3, where the p_T spectrum is shown separately for the two production processes using the CTEQ PDF.

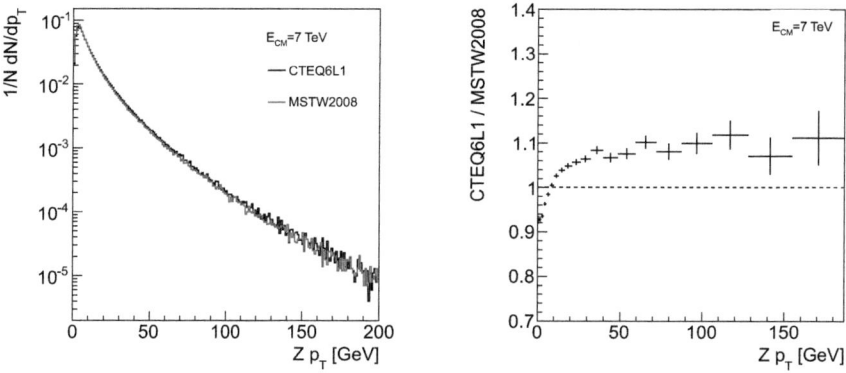

Figure 9.2: Transverse momentum spectrum with the MSTW2008 and the CTEQ6L1 PDF sets for an inclusive $Z \to \mu\mu$ event sample and the corresponding ratio.

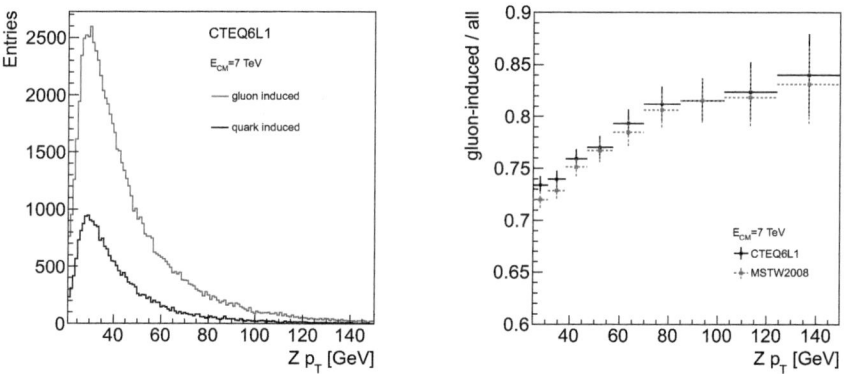

Figure 9.3: Transverse momentum spectrum in $Z \to \mu\mu$ events with CTEQ6L1 for the gluon induced and the quark induced production processes and the ratio of the two different production processes with CTEQ6L1 and MSTW2008 PDF sets.

Overall, the gluon-induced process is roughly a factor of 3 more frequent and very similar for the MSTW based spectrum. The fraction of the gluon-induced events is shown for

CTEQ and MSTW, and a slight deviation towards lower values is observed for MSTW compared to CTEQ, especially in the low p_T region. The p_T dependance is comparable for both PDFs. The high p_T region is dominated by the gluon-induced processes, while at lower values it decreases. The gluon contribution rises with increasing p_T, from slightly above 70% at 30 GeV to close to 85% above 100 GeV. At very low values the calculation diverges and different settings have to be used[1]. Consequently, the excess when using CTEQ compared to MSTW in fig. 9.2 can be understood as due to the larger fraction of gluon-induced processes.

To illustrate the magnitude of a change in the composition, fig. 9.4 shows the p_T spectrum for $W \to \mu\nu$ events produced with the CTEQ PDF after varying the gluon fraction by $\pm 5\%$.

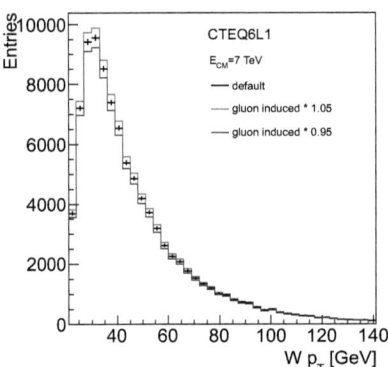
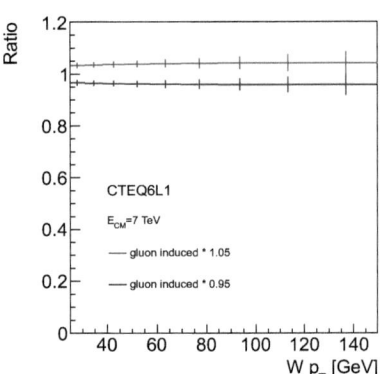

Figure 9.4: Transverse momentum spectrum in $W \to \mu\nu$ events with CTEQ6L1 for a variation of the gluon-induced contribution of $\pm 5\%$ compared to the default. On the right, the ratio of the varied to the default spectrum.

As expected and in agreement with the figures seen before, the spectrum is shifted to higher p_T values with the increase of the gluon contribution and correspondingly in the opposite case.

Therefore, the composition of the production processes has a, even though small, measurable influence on the shape of the p_T spectrum and can be used to further constrain and improve the gluon PDF.

---

[1]For these studies the MC settings were such that the high p_T would be described and consequently the figures are shown above a certain cut-off, here chosen to 30 GeV.

9.2 Transverse Momentum Measurement

The determination of the transverse momentum spectrum of the W boson

$$p_T^W = \sqrt{(p_x^\mu + p_x^\nu)^2 + (p_y^\mu + p_y^\nu)^2}$$

requires the measurement of the missing transverse energy, which is based on the hadron and jet activity in the event. The Z boson on the other hand is reconstructed via the Z four-vector from the two muons, which are measured with great precision. Figure 9.5 compares the transverse momentum for Z and W events as generated with MC simulations and after the reconstruction chain, and the p_T resolution defined as the difference between the generated and the reconstructed p_T normalized to the generated p_T. The resolution of the Z boson is excellent with a mean value of -0.0034 and a RMS of 0.114. It is roughly a factor of 10 better than for the W, which has a mean of -0.573 and a RMS of 1.156. We observe an asymmetry in the resolution distribution of the W, which visualizes the broadening of the W p_T spectrum after reconstruction with respect to the true distribution. As the underlying QCD dynamics is the same for W and Z events, the Z p_T can be utilized to control the W p_T spectrum measured in data. Namely, the W and Z events can be separated in events with different jet contribution. As was detailed in section 2.2.3, in events with jets, the boson p_T and the jet p_T have to balance each other. If more than one jet is produced, the vector sum of the measured jets balances the boson p_T. Thus, the Z p_T can be measured in two different ways, via the precisely measured lepton four-vector or via the reconstructed jet. The W p_T distribution can then be approximated by the jet p_T, which is controlled by the measured Z p_T. The W boson cross section is roughly an order of magnitude larger compared to the Z boson. The reach of the W is correspondingly larger and can be exploited in the high p_T region, were the statistical error becomes important.

In conclusion, W measurements are used to study the high p_T sector, utilizing a data-driven control of the measurement. The control is given by the well measured Z boson. The low p_T sector can be accessed with the Z boson, which gives precise measurements to very low values.

Once all the distributions are understood and under control, the measured p_T spectrum in data can be translated back to the generator level via a correction factor determined from MC. Figure 9.6 shows the correction factor defined as the ratio of the reconstructed over the generated W p_T distribution as a function of p_T. The large correction factor below 50 GeV reflects the jet cut-off due to the divergences in the low p_T region and should not

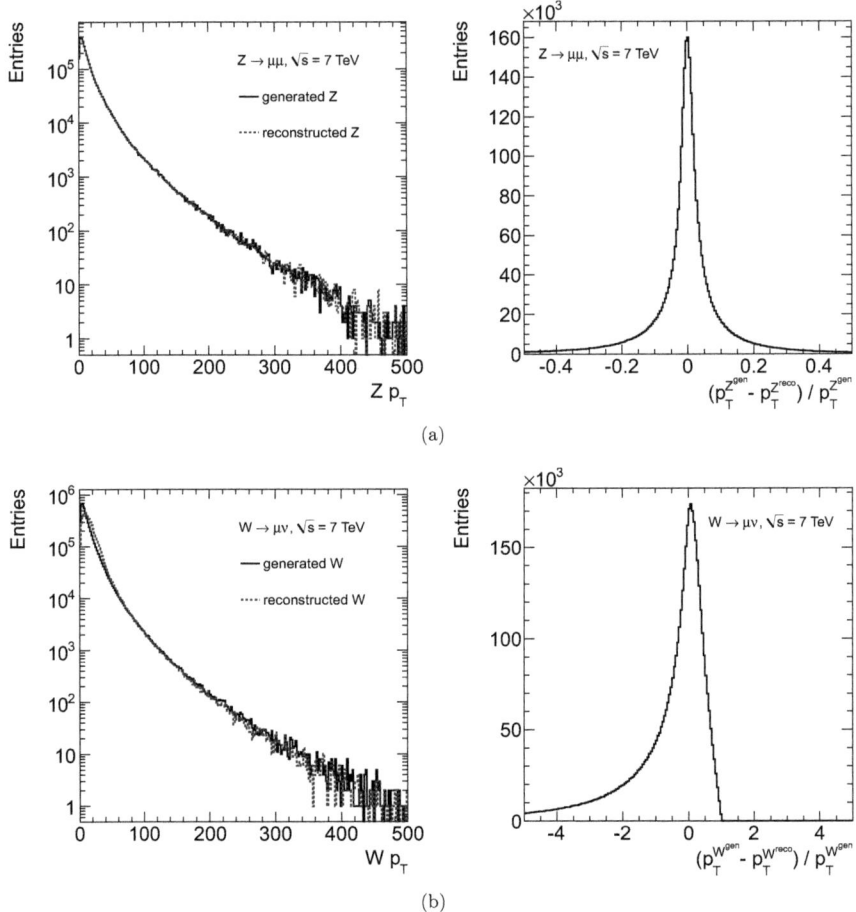

Figure 9.5: Transverse momentum at generator and reconstruction level and the p_T resolution for Z (a) and W (b) bosons as expected from simulation.

be trusted. Above, the corrections are reasonably stable.

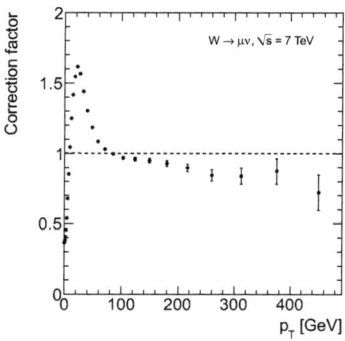

Figure 9.6: Correction factor for the W transverse momentum spectrum calculated from the ratio of the MC p_T prediction at the stable particle level over the detector level.

In the following, the measurements of the p_T spectrum are presented for the 2011 data sample corresponding to an integrated luminosity of $1.5\,\text{fb}^{-1}$. For the signal simulation a MadGraph MC sample is used with the CTEQ6L1 PDF set (section 2.3).

9.2.1 Results for Z p_T

In the following, Z p_T denotes the transverse momentum measured from the four-vector of the two selected muons and jet p_T refers to the balancing p_T determined from the reconstructed jets, which particularly means a cut on $p_T > 30\,\text{GeV}$.

Figure 9.7 shows the inclusive Z p_T distribution in a linear and a logarithmic scale for data and simulation and the ratio of data over MC. The data are well described by the MC in the intermediate region, roughly above 10 GeV and below 100 GeV. Above, the discrepancy is larger, but also the statistical error increases significantly. Below roughly 10 GeV a discrepancy between data and MC becomes evident (fig. 9.8). This is the region which is dominated by soft gluon radiation and perturbative calculations can not be applied anymore. MC simulations rely on parametrized models and parton showering. The measurements can therefore be used to improve the simulations.

In fig. 9.9 the Z events are divided according to the number of reconstructed jets in the event. Figure 9.9a shows the jet multiplicity for all selected Z events; in figs. 9.9b–9.9d the p_T spectrum for events with no, one and two jets is shown. Similar to the inclusive case, the agreement is good in the intermediate region and starts to disagree for smaller p_T. Figure 9.10 shows the low p_T region for the Z transverse momentum in events with one or two jets. The inclusive distribution is dominated by the events were no additional jet is found and looks practically identical. For the events with one or two jets the discrepancy between data and MC in the low p_T region extends to higher values than in the inclusive case. Contrary to the inclusive case in the Z + 1 jet events the data point lie below MC for a p_T up to roughly 20 GeV.

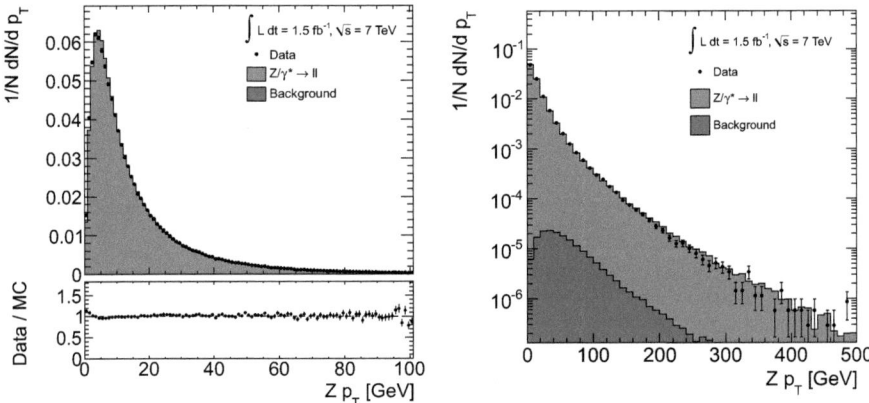

Figure 9.7: Z transverse momentum distribution for $1.5\,\text{fb}^{-1}$ data collected in 2011 compared to signal and background simulation in linear (left) and logarithmic (right) scale.

In fig. 9.11 the transverse momentum measured with the balancing jet is shown. In the events with 2 additional jets the vector sum of the two jets is built and the shown p_T spectrum is equivalent to one large jet balancing the Z p_T. The dip around 50 GeV can be ascribed to the p_T cut-off of 30 GeV to reconstruct a jet. The data is in good agreement with the MC simulation.

Finally, in fig. 9.12 the Z and jet p_T spectra are shown overlaid for data and compared to Z p_T from simulation. The Z p_T distribution contains events with no, one or two reconstructed jets; the jet p_T distribution events with one or two jets. At the lowest bin a notable discrepancy is observed, It is interpreted as an edge effect from the jet p_T cut-off of 30 GeV and the distribution should be used only well above from the cut-off.

W and Z Transverse Momentum Spectrum

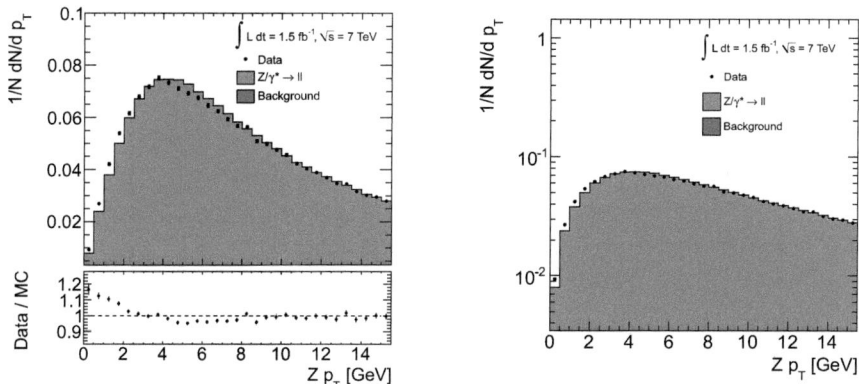

Figure 9.8: Z transverse momentum distribution in the low p_T region for data and MC in linear (left) and logarithmic (right) scale.

In conclusion, we have seen that the Z p_T spectrum, precisely measured via the four-vector of the two muons, agrees well with MC in the inclusive case and in the events with 0, 1 and 2 jets, except for the very low p_T regions. The measurements with p_T smaller than 10 GeV can be used to improve the MC simulations. The jet in Z + 1 jet events, or the vectorial sum of the jets in Z + 2 jet events, balances the Z p_T as expected from simulation, and can thus be used as a control of the W p_T.

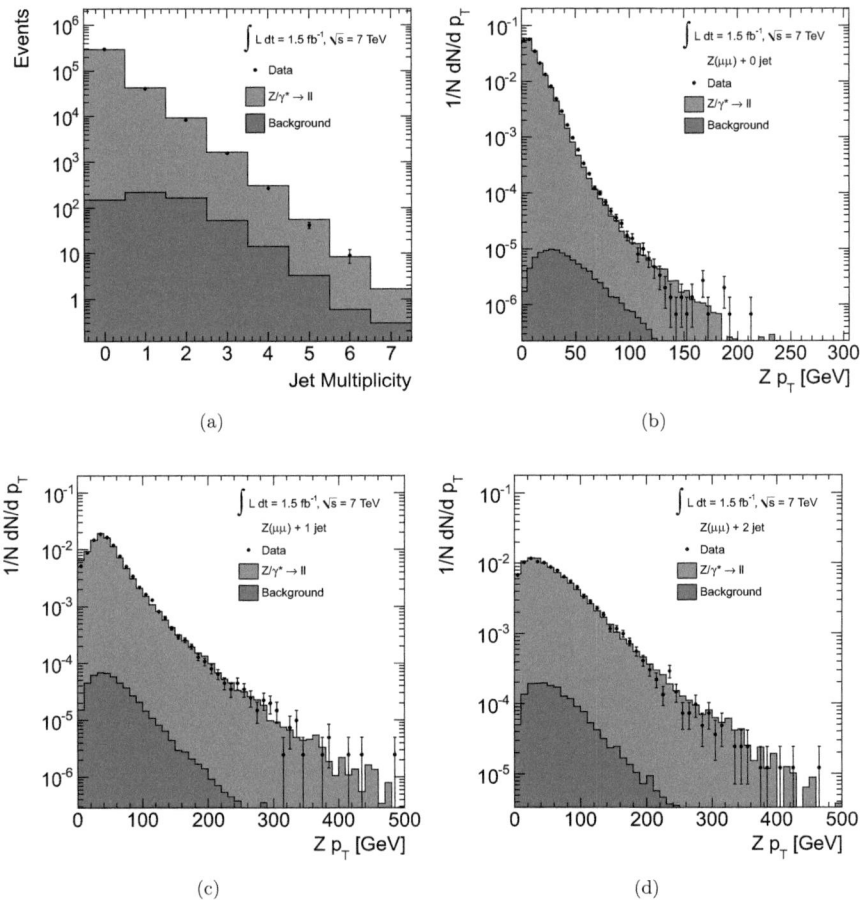

Figure 9.9: Jet multiplicity (a) and Z transverse momentum distribution for $1.5\,\text{fb}^{-1}$ data collected in 2011 compared to signal and background simulation for events with no (b), one (c) and two (d) reconstructed jets.

W and Z Transverse Momentum Spectrum

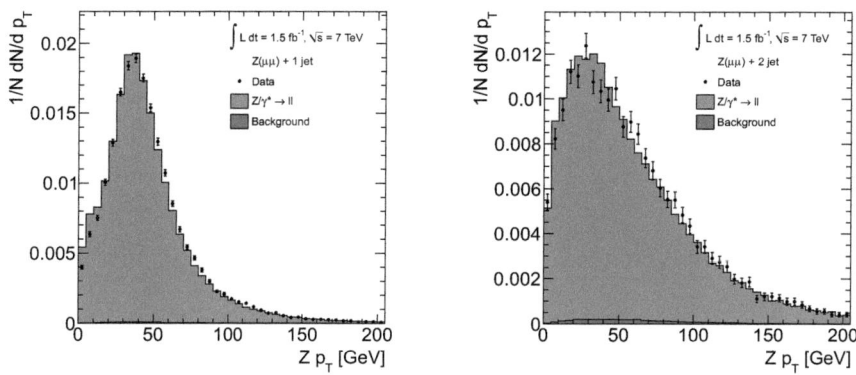

Figure 9.10: Z transverse momentum distribution in the low p_T region for data and MC for events with one (left) and two (right) reconstructed jets.

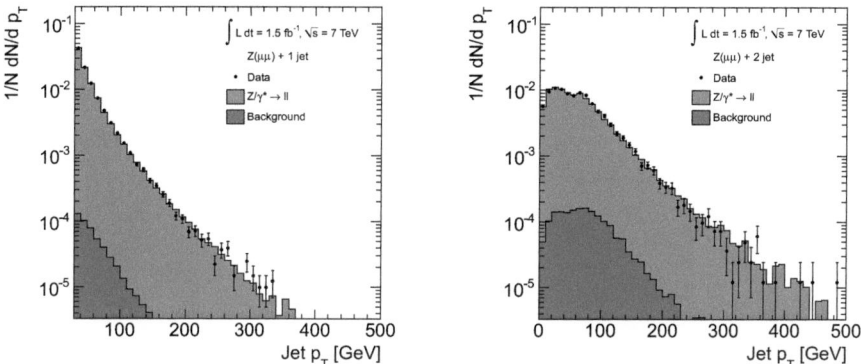

Figure 9.11: Jet transverse momentum distribution for data and MC for Z events with one (left) and two (right) reconstructed jets. The p_T for the two jets is built from the vector sum of the two jets, equivalent to one large jet.

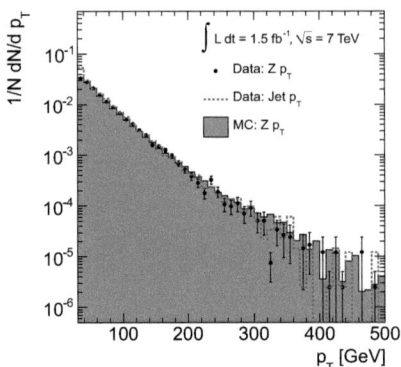 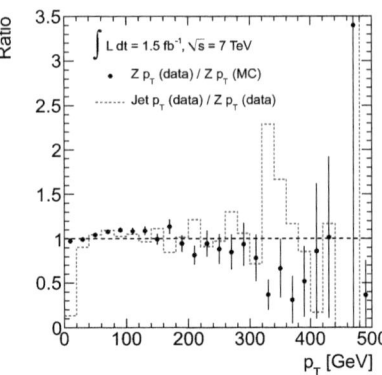

Figure 9.12: Z and jet transverse momentum measured in data compared to the simulated Z p_T spectrum. The Z p_T distribution contains events with no, one or two reconstructed jets; the jet p_T distribution events with one or two jets. Note the different binning in the ratio plot.

9.2.2 Results for W p_T in Events with One and Two Jets

The transverse momentum of the W boson can be measured through its two decay products, the muon and the neutrino, or via the balancing jet in events with one or more jets. In contrast to the Z p_T, the first method delivers similar results as the measurement of the balancing jet, due to the escaping neutrino leaving as trace only missing E_T.[2] Here one can take advantage of the precisely measured Z p_T, which serves as control and validation of the jet p_T, and which should be identical for the two event types. In fig. 9.13 the p_T distribution is shown measured via the μ-MET system for data compared to signal and background simulation in linear and logarithmic scale. Data deviates from MC below roughly 15 GeV, which can be expected. The distribution measured via the jet p_T is shown in fig. 9.14 in linear and logarithmic scale. A slight broadening of the shape measured with data can be observed compared to MC. The jet p_T in W events with two reconstructed jets is shown in fig. 9.15, where the p_T is built from the vector sum of the two jets. We observe a similar dip around 50 GeV as found in the Z + 2 jets events, which is interpreted as a jet-cutoff effect. Additionally, a slight excess of data with higher p_T is observed compared to MC in the region up to roughly 140 GeV.

In fig. 9.16 the transverse momentum spectrum reconstructed through the μ-MET system and via the jet is shown for W + 1 jet events measured in data and compared to the

---
[2] Note the same principle in the reconstruction of jets and MET

W and Z Transverse Momentum Spectrum

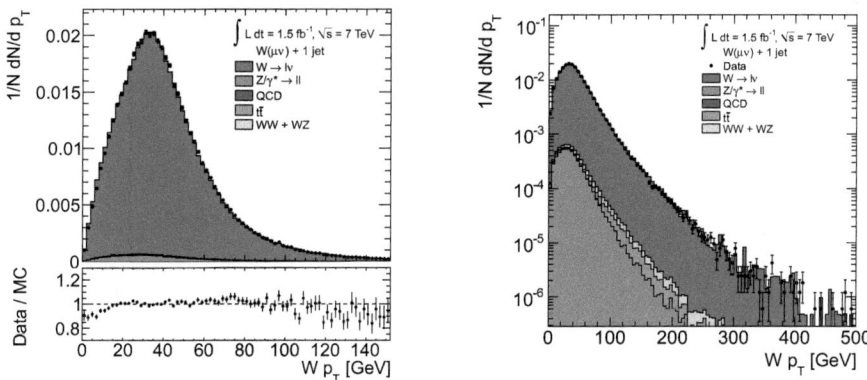

Figure 9.13: W transverse momentum distribution measured via the μ-MET system in W + 1 jet events using 1.5 fb^{-1} data collected in 2011 compared to signal and background simulation in linear (left) and logarithmic (right) scale.

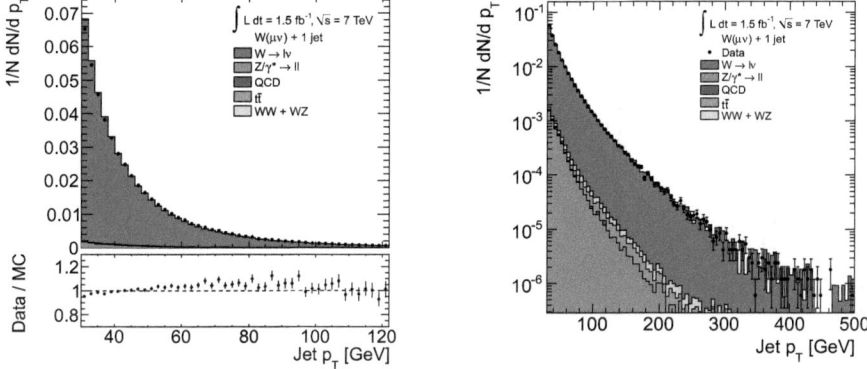

Figure 9.14: W transverse momentum distribution measured via the reconstructed jet in W + 1 jet events compared to signal and background simulation in linear (left) and logarithmic (right) scale.

simulated (μ-MET) p_T spectrum. Again we observe a cut-off effect for low p_T values. Above that region, the agreement between the reconstructed jet p_T and the p_T from the (μ-MET) system is good.

For the W p_T measurement we can conclude, that the agreement between data and MC with the CTEQ6L1 PDF set in events with one jet present is remarkably good above the cut-off effect from the jet p_T. The agreement is such, that no relevant difference can be

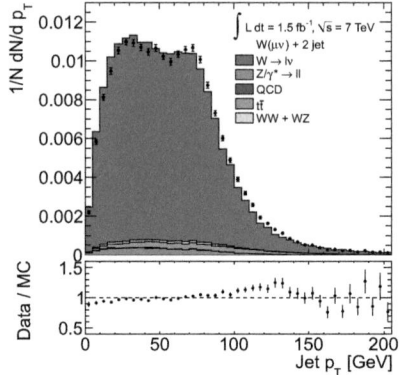 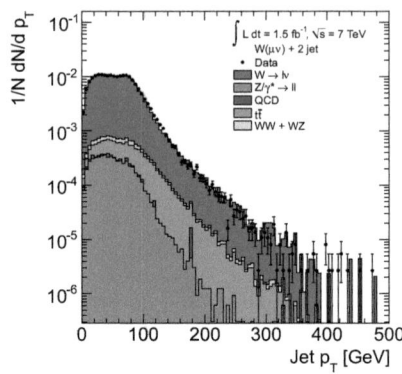

Figure 9.15: W transverse momentum distribution measured through the reconstructed jet in W + 2 jet events compared to signal and background simulation in linear (left) and logarithmic (right) scale. The p_T is built from the vector sum of the two jets.

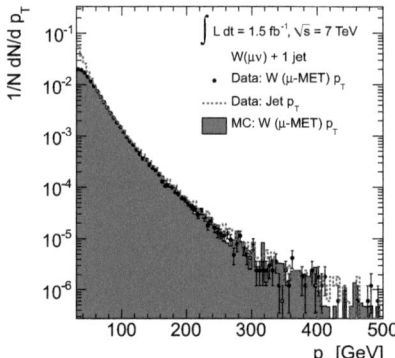 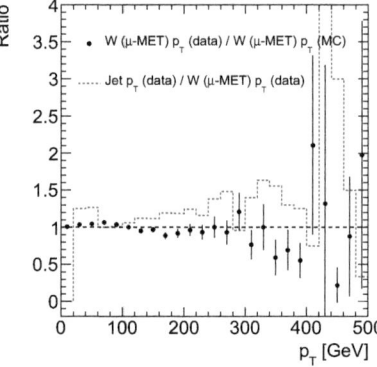

Figure 9.16: Transverse momentum spectrum reconstructed through the μ-MET system (black dots) and via the jet (red line) for W + 1 jet events measured in data and compared to the simulated W p_T spectrum. Note the different binning in the ratio plot.

observed between the measurement via the jet and the μ-MET system. The Z p_T could be used to validate the utilization of the jet p_T for the W p_T without any corrections. In W events with 2 jets present, a slight excess of data with high p_T is observed in the region up to roughly 140 GeV. The excess is not visible in the Z events, however, the statistical significance is less there. This particular feature could be a reflection of the jet cut-off and thus, the events with 2 or more jets should be studied in more detail with a larger dataset to verify this hypothesis. Above and within the errors, the agreement between data and MC is very good.

Chapter 10

Conclusions

In this thesis, measurements of the processes pp $\to$ WX $\to$ $\mu\nu$X and pp $\to$ ZX $\to$ $\mu\mu$X with CMS data at a center of mass energy of $\sqrt{s} = 7\,\text{TeV}$ were presented. A robust and clean signal selection was developed for all measurements. The muon as well as the W and Z selections were studied prior to the data taking period with MC simulations, and validated later with data. A remarkably good agreement was found for the selection variables between data and MC.

The selected W and Z events with one primary vertex were analyzed in detail, with the aim to gain a better understanding of hard diffractive processes. The measurements show that the tuning of the models describing W(Z) production in combination with multiparton interactions with and without a diffractive component in the entire CMS phase space needs substantial modifications. The measurements of the forward energy flow and the central charged-particle multiplicities presented here can be used for an improvement of these models. For the energy flow measurement, the hadron level corrections were determined. This allows a direct comparison of the CMS measurements with theory and other experiments.

Diffractive events with a hard scale present have been found in previous experiments. Here, it was shown that diffractive events with W or Z bosons are produced at the LHC. The commonly used Large Rapidity Gap signature did not provide enough evidence due to the lack of reliable MC models. A diffractive component in the LRG events could be confirmed with the use of an asymmetry variable, showing that the majority of the leptons are found in the hemisphere opposite to the gap. Fitting the available MC simulations to data, it was found that the diffractive fraction of the data sample with a LRG is $(50.0 \pm 9.3\,(\text{stat.}) \pm 5.2\,(\text{syst.}))\%$.

Finally, the measurements of the transverse momentum of W and Z bosons were presented. It was shown that the Z p_T agrees well with the MC expectations for large transverse momentum, while for the region below roughly 10 GeV a discrepancy is observed. The low p_T region is dominated by soft gluon radiation and parametrized models have to be used to describe it. The p_T measurements can be used as input for these models.

Above 30 GeV, the Z p_T is used to validate and control the jet p_T spectrum, which is used to measure the W p_T. This is done by exploiting momentum conservation in W and Z production with 1 or 2 jets, where the jet in W(Z) + 1 jet events, or the vector sum of the jets in W(Z) + 2 jet events, recoils against the W(Z) boson. The W p_T spectrum is compared to the predictions of MC simulations with the CTEQ6L1 PDF sets. Good agreement was found well above the cut-off effect from the jet p_T.

The presented work could eventually be carried out with a larger data sample to further increase the precision of the measurement. With an order of magnitude more data in low pileup conditions, a very rich diffractive physics program could be done. The transverse momentum measurement could be repeated with the larger 2012 data sample and ultimately at a center-of-mass energy of 14 TeV.

Appendix A

Diffractive Processes: Additional Event Displays

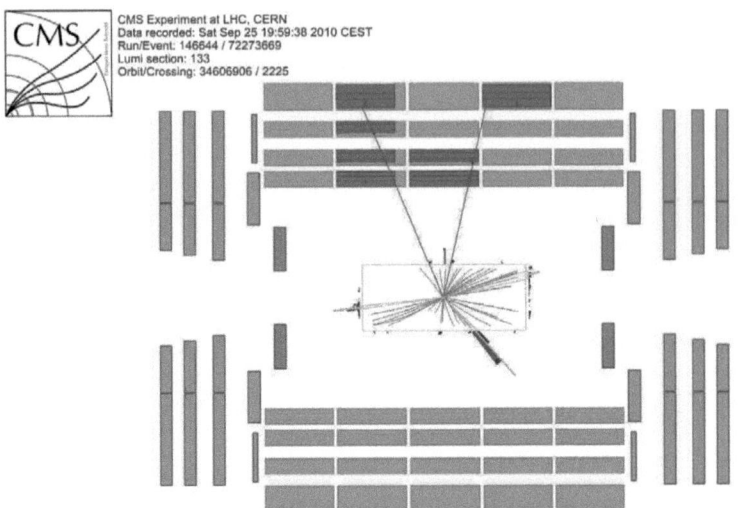

(a) r-z view with the positive pseudorapidity on the right side

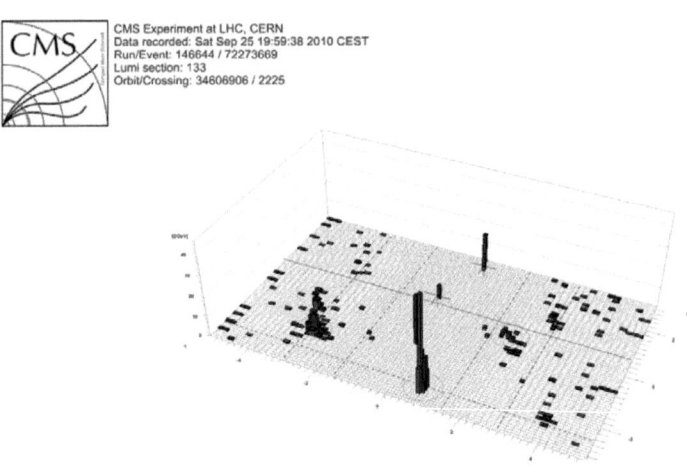

(b) 3 dimensional view of the detector

Figure A.1: Event Displays for a $Z \to \mu\mu$ candidate without large pseudorapidity gap for two different perspectives visualizing the particle deposits in the detector.

Diffractive Processes: Additional Event Displays

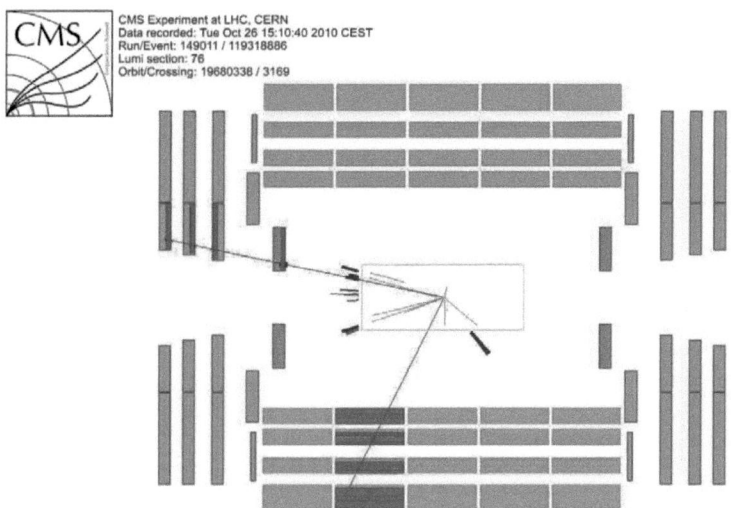

(a) r-z view with the positive pseudorapidity on the right side

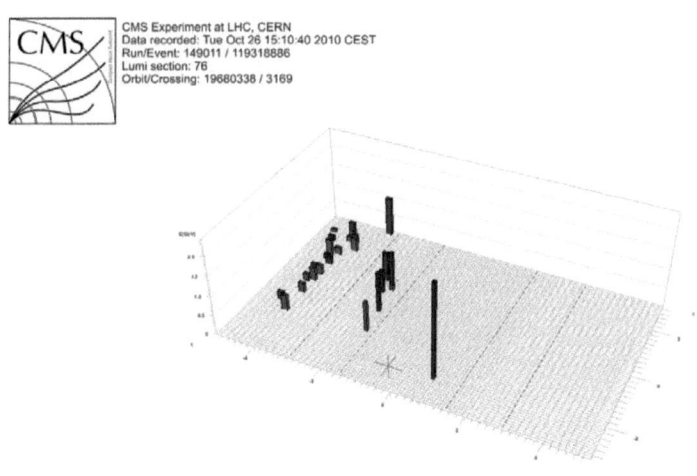

(b) 3 dimensional view of the detector

Figure A.2: Event Displays for a $Z \to \mu\mu$ candidate with a large pseudorapidity gap on the positive detector hemisphere for two different perspectives visualizing the particle deposits with a large gap in the detector.

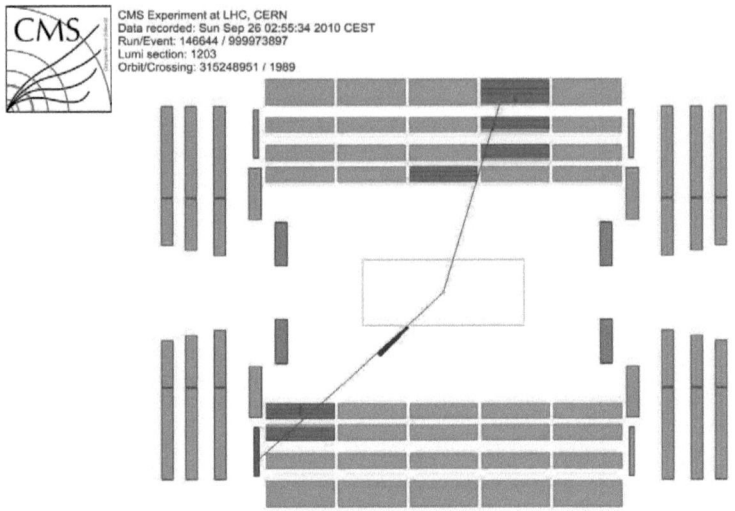

(a) r-z view with the positive pseudorapidity on the right side

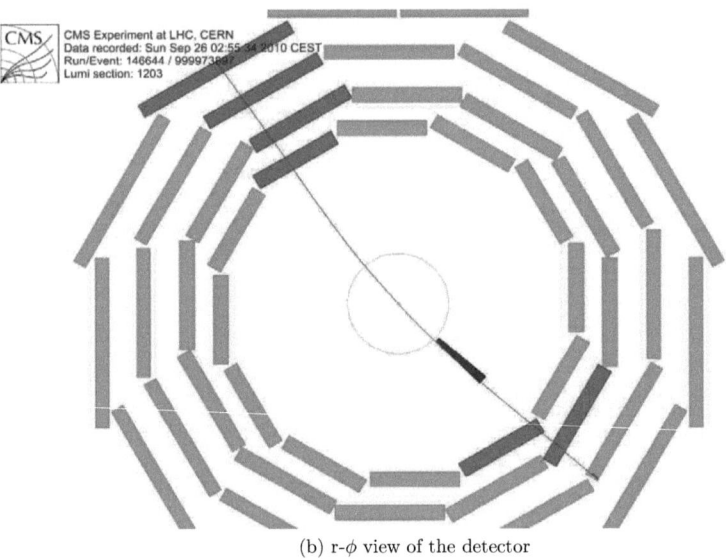

(b) r-ϕ view of the detector

Figure A.3: Event Displays for a $Z \to \mu\mu$ candidate with zero energy in both HF calorimeters for two different perspectives visualizing the muon hits in the detector.

Appendix B

Forward Energy Flow and Central Charged-Particle Multiplicity: $Z \to \mu^+\mu^-$ distributions

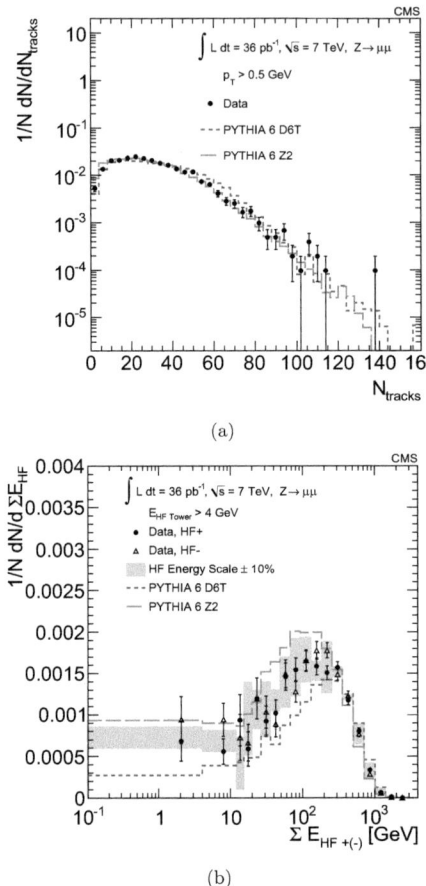

Figure B.1: The charged-particle multiplicity (a) and the summed HF+ and HF− energy (b) distributions for $Z \to \mu\mu X$ candidate events are shown for data and MC simulations, including pileup, with different tunes for the underlying event. The band shown for the HF energy distributions indicates the uncertainty related to a $\pm$ 10% HF energy scale variation.

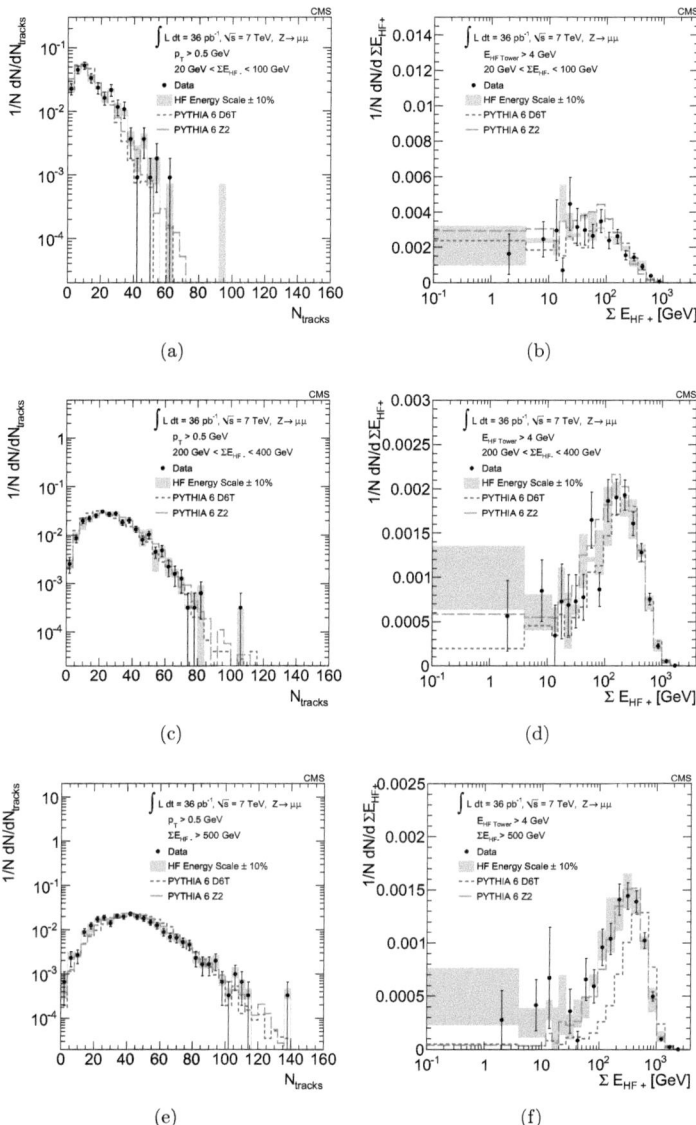

Figure B.2: The charged-particle multiplicity and the summed HF+ energy distributions in the data and from MC simulations with different tunes, for the three HF− energy intervals of (a) and (b) 20-100 GeV, (c) and (d) 200-400 GeV, and (e) and (f) > 500 GeV.

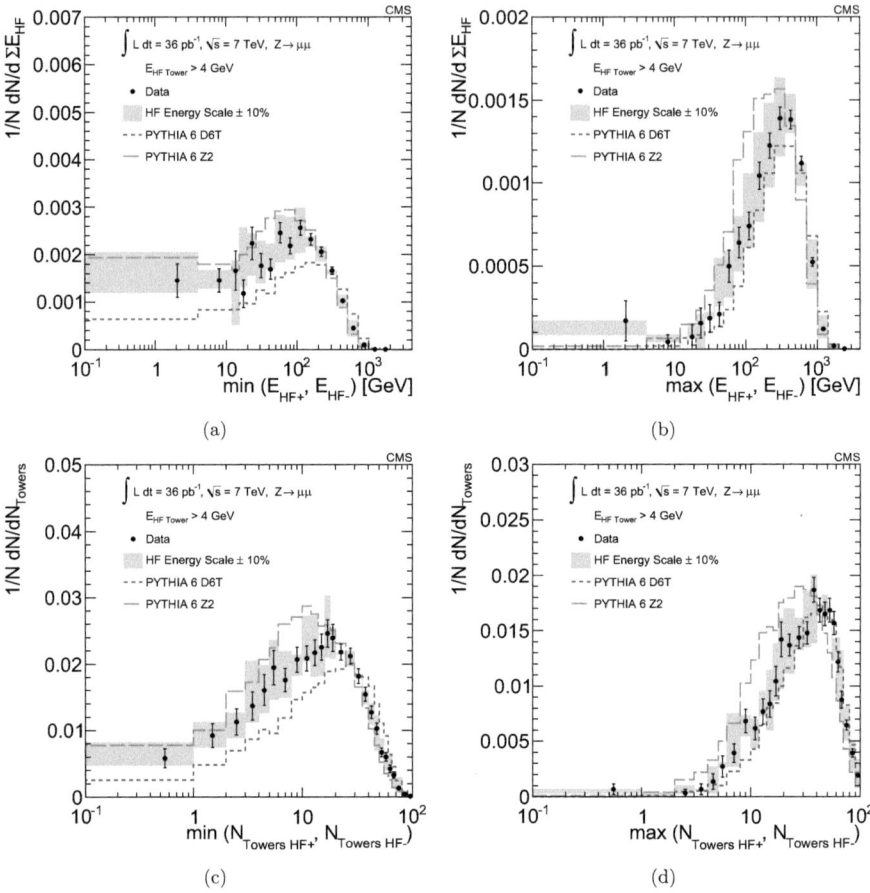

Figure B.3: HF energy and tower multiplicity distributions in $Z \to \mu\mu X$ events for data and different MC tunes. (a) and (c) show the minimum (min (E_{HF+}, E_{HF-}) and min ($N_{TowerHF+}$, $N_{TowerHF-}$, respectively) and (b) the maximum (max (E_{HF+}, E_{HF-} and max ($N_{TowerHF+}$, $N_{TowerHF-}$, respectively) of the energy depositions and the tower multiplicities per event in the HF+ and HF− calorimeters.

Appendix C

W and Z Transverse Momentum: Ratios in Events with One Jet

The production process of W and Z bosons are in principle equivalent. The main difference in the production properties comes from the different masses, but some large theoretical and experimental uncertainties, such as luminosity or radiative corrections, are the same for both processes. Therefore, by measuring the ratio of the two distributions, these errors are canceled and a much more precise measurement is achieved, which can be compared to a more precise prediction. As a short outlook of the p_T measurements, the ratio of the W^+ and W^- distributions (shown in fig. C.1), W^+/W^- is shown in fig. C.2 and the ratio of the W and Z distributions, W/Z in fig. C.3.

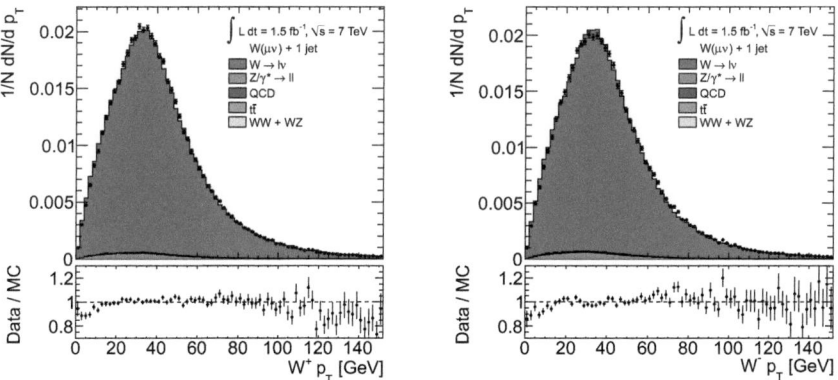

Figure C.1: Transverse momentum spectrum for data and MC for W + 1 jet events for W^+ events (left) and W^- events (right). The transverse momentum is obtained from the ($\mu^\pm$-MET) system.

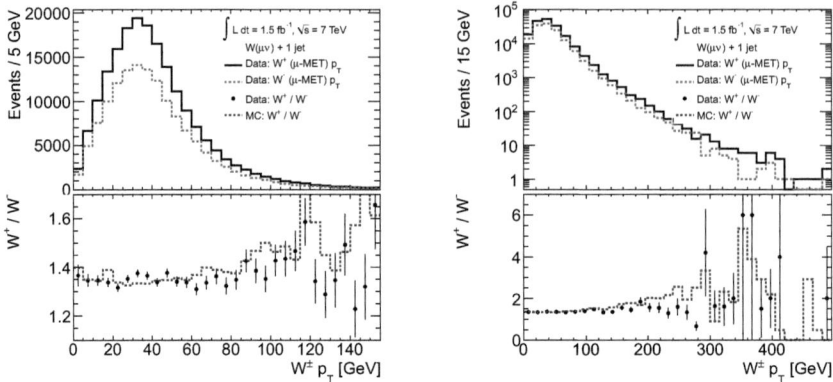

Figure C.2: Transverse momentum for W^+ and W^- events with one jet and the corresponding ratio W^+/W^- in linear (left) and logarithmic (right) scale. The dashed blue line indicates the expected ratio from MC. The transverse momentum is obtained from the ($\mu^\pm$-MET) system.

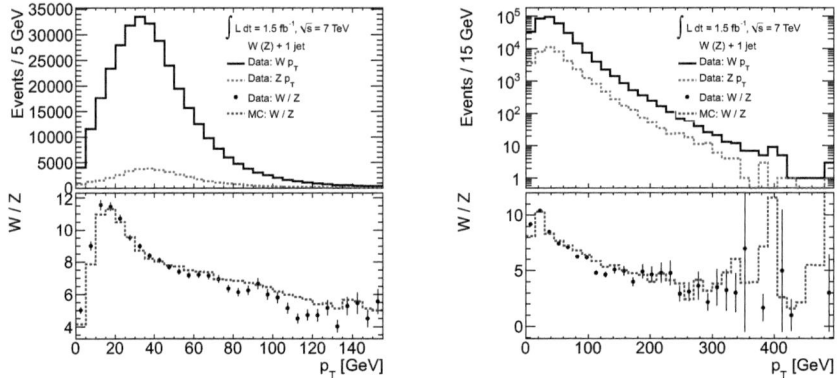

Figure C.3: Transverse momentum for W and Z events with one jet and the corresponding ratio W/Z in linear (left) and logarithmic (right) scale. The dashed blue line indicates the expected ratio from MC. The transverse momentum for the Z is obtained from the dimuon system and for the W from the ($\mu^\pm$-MET) system.

List of Tables

2.1	Building blocks of the Standard Model of Particle Physics	4
2.2	Monte Carlo datasets for signal and background.	23
4.1	Fraction of good and fake muons for various muon selection cuts.	56
5.1	Number of events for each Z selection cut.	66
5.2	Number of Z events with one, two, or more jets.	66
5.3	Number of events for each W selection cut.	73
5.4	Number of W events with one, two, or more jets.	74
6.1	Number of W and Z candidate events with a single primary vertex	90
7.1	Mean energy depositions and tower multiplicities for W events.	95
7.2	Pseudorapidity ranges for the five HF detector rings.	105
8.1	Number of W and Z LRG events before PU corrections.	114
8.2	Number of W and Z LRG events after PU corrections.	115

List of Figures

2.1 Schematic view of the W/Z production in proton-proton collisions. 12
2.2 PDFs for the LHC at NLO for two Q^2 values from MSTW. 13
2.3 Leading order feynman diagrams for W^+, W^- and Z^0 production. 14
2.4 Q^2 vs. x for the LHC at $\sqrt{s} = 7\,\text{TeV}$. 15
2.5 Feynman diagrams for the first order W + jet production. 17
2.6 Diagrams for the various processes at the LHC. 20
2.7 Sketch for the various diffractive processes at the LHC. 21

3.1 Cross sections for different processes as a function of $\sqrt{s}$. 26
3.2 Overview of the CERN acceleration complex. 28
3.3 Schematic view of the Compact Muon Solenoid. 30
3.4 Slice of CMS transverse to the beam axis. 31
3.5 The CMS tracking system in the $r - z$ plane. 33
3.6 Layout of the CMS ECAL detector. 34
3.7 ECAL energy resolution as a function of the reconstructed energy. 35
3.8 Schematic view of the transverse segmentation of the HF towers. 37
3.9 Schematic view of one quarter of the CMS muon system. 38
3.10 Layout of a DT and a CSC chamber in the muon system. 39
3.11 Resonances in the dimuon mass spectrum measured with 2010 CMS data. . 42

4.1 Muon η and p_T for three muon reconstruction algorithms. 50
4.2 Muon reconstruction efficiency as a function of η. 51
4.3 Muon reconstruction efficiency as a function of p_T. 51
4.4 Muon momentum resolution as a function of p. 52
4.5 Muon momentum resolution as a function of eta. 53
4.6 Sketch of the isolation cone for the muon isolation algorithm. 54
4.7 Muon cut efficiency for four different isolation variables. 55
4.8 Muon selection variables for good and fake muons. 57

LIST OF FIGURES

5.1	Generated MET and two different reconstruction algorithms for W events.	61
5.2	MET distributions for three jet types for W events from data.	62
5.3	Jet energy scale uncertainty as a function of the jet p_T for three jet types. .	63
5.4	Muon selection variables for Z events.	67
5.5	Muon ϕ and η distributions for Z events.	68
5.6	Lower and higher p_T distributions for the two Z muons.	70
5.7	Dimuon invariant mass for selected Z events.	71
5.8	Dimuon invariant mass for selected Z events with no, one, and two jets. . .	72
5.9	Muon selection variables for W events.	75
5.10	Muon ϕ and η distributions for W events.	76
5.11	Muon p_T and MET distributions for W events before the M_T cut.	78
5.12	Muon p_T and MET distributions for W events after the M_T cut.	79
5.13	Transverse mass for selected W events.	80
5.14	Transverse mass for selected W events with no, one, and two jets.	81
6.1	Event Displays for a $W \to \mu\nu$ candidate with a LRG.	86
6.2	Sketches of the process pp $\to$ W(Z)X with MPI and diffractive component.	87
6.3	Number of reconstructed vertices for three data-taking periods.	88
6.4	Distance between the W-vertex and an additional vertex.	89
7.1	Charged-particle multiplicities and p_T spectrum for W events.	93
7.2	HF+ and HF− energy distribution for W events.	94
7.3	Charged-particle multiplicities for three different HF− energies.	98
7.4	HF+ energy distributions for three different HF− energies.	99
7.5	Correlations of HF+ and HF− distributions.	100
7.6	Minimum and maximum HF energy and tower multiplicity.	101
7.7	Generator vs. detector level HF energy distribution.	104
7.8	Generator vs. detector level HF energy distribution for five η bins.	106
7.9	Generator vs. detector level HF energy distribution for a constrained η acceptance.	107
7.10	Generated and reconstructed energy flow as a function of η.	108
7.11	Correction factors for five η-bins.	109
7.12	Response matrix for the forward energy flow for two tunes.	110
7.13	Response matrix in a two-dimensional plot.	110
7.14	Response matrix in a lego plot.	111
8.1	$\tilde{\eta}$ distributions for W events.	118
8.2	Charged-particle multiplicity and HF energy distributions in LRG events. .	119

LIST OF FIGURES

8.3   Diffractive parton distribution function vs. conventional proton PDF. . . . 120
8.4   Signed lepton pseudorapidity distribution in W events with a LRG. 121

9.1   Fractional uncertainties of the parton-parton luminosity functions. 124
9.2   Z p_T spectrum with two different PDFs. 125
9.3   Z p_T for the gluon induced and the quark induced production processes. . . 125
9.4   W p_T for a variation of the gluon-induced contribution of $\pm 5\,\%$. 126
9.5   Generator and reconstruction level p_T spectrum and resolution for W and Z events. 128
9.6   Correction factor for the W transverse momentum spectrum 129
9.7   Z p_T spectrum for $1.5\,\mathrm{fb}^{-1}$ data. 130
9.8   Z p_T spectrum for the low p_T region. 131
9.9   Jet multiplicity and Z p_T spectrum in Z + 0, 1 and 2 jet events. 132
9.10 Z p_T for the low p_T region in Z + 1 and 2 jet events. 133
9.11 Jet p_T in Z + 1 and 2 jet events. 133
9.12 Z and jet transverse momentum measured in data compared to simulation. 134
9.13 W p_T measured via the μ-MET system in W + 1 jet events. 135
9.14 W p_T measured via the Jet p_T in W + 1 jet events. 135
9.15 W p_T measured via the Jet p_T in W + 2 jet events. 136
9.16 W and jet transverse momentum measured in data compared to simulation. 136

A.1   Event Displays for a $Z \to \mu\mu$ candidate without LRG. 142
A.2   Event Displays for a $Z \to \mu\mu$ candidate with a LRG in HF$^+$. 143
A.3   Event Displays for a $Z \to \mu\mu$ candidate with a LRG in both HF. 144

B.1   Charged-particle multiplicity and HF energy distributions for Z events. . . 146
B.2   Charged-particle multiplicity and HF+ energy distributions for three HF− energy intervals for Z events. 147
B.3   Minimum and maximum HF energy and tower multiplicity for Z events . . 148

C.1   W$^+$ and W$^-$ transverse momentum spectra in events with one jet. 149
C.2   W$^+$/W$^-$ transverse momentum spectrum in events with one jet. 150
C.3   W/Z transverse momentum spectrum in events with one jet. 150

Bibliography

[1] CMS Collaboration, "Forward Energy Flow, Central Charged-Particle Multiplicities, and Psuedorapidity Gaps in W and Z Boson Events from pp Collisions at $\sqrt{s} = 7$ TeV.", *Eur. Phys. J. C* **72** (2012) 1839. doi:`10.1140/epjc/s10052-011-1839-3`.

[2] K. N. at al. (Particle Data Group), "The Review of Particle Physics", *J. Phys. G* **37** (2010) 075021.

[3] S. Glashow, "Elementary Particle Theory", *Nucl. Phys.* **20** (1961) 579.

[4] S. Weinberg, "A Model of Leptons", *Phys. Rev. Lett.* **19** (1967) 1264.

[5] A. Salam, "Elementary Particle Physics", *N. Svartholm (Almqvist and Wiksell, Stockholm)* (1968).

[6] The Gargamelle Collaboration, "Search for Elastic Muon Neutrino Electron Scattering", *Phys. Lett. B* **46** (1973) 121.

[7] The UA1 Collaboration, "Experimental observation of lepton pairs of invariant mass around 95 GeV at the CERN SPS collider.", *Phys. Lett. B* **126** (1983) 398.

[8] The UA2 Collaboration, "Evidence for $Z^0 \to e^+e^-$ at the CERN $p\bar{p}$ collider.", *Phys. Lett. B* **129** (1983) 130.

[9] The UA1 Collaboration, "Experimental Observation of Isolated Large Transverse Energy Electrons with Associated Missing Energy at $\sqrt{s} = 540$ GeV.", *Phys. Lett. B* **122** (1983) 103.

[10] The UA2 Collaboration, "Observation of single isolated electrons of high transverse momentum in events with missing transverse energy at the CERN pp collider", *Phys. Lett. B* **122** (1983) 476.

[11] M. E. Peskin and D. V. Schroeder, "An Introduction to Quantum Field Theory". Westview Press, Colorado, 1995.

[12] G. Dissertori, I. Knowles, and M. Schmelling, "Quantum Chromodynamics". Oxford Science Publications, 2003.

[13] F. Halzen and A. Martin, "Quarks and Leptons: An Introductory Course in Modern Particle Physics". John Wiley & Sons, 1984.

[14] J. C. Collins, D. E. Soper, and G. Sterman, "Factorization of Hard Processes in QCD", *Adv. Ser. Direct. High Energy Phys.* **5** (1988) 1.

[15] R. Feynman, "Very High-Energy Collisions of Hadrons.", *Phys. Rev. Lett.* **23** (1969) 1415.

[16] Y. L. Dokshitzer, "Calculation of the Structure Functions for Deep Inelastic Scattering and e^+e^- Annihilation by perturbation Theory in Quantum Chromodynamics", *Sov. Phys. JETP* **46** (1977) 641.

[17] V. Gribov and L. Lipatov, "Deep Inelastic ep Scattering in Perturbation Theory", *Sov. J. Nucl. Phys.* **15** (1972) 438.

[18] G. Altarelli and G. Parisi, "Asymptotic Freedom in Parton Language", *Nucl. Phys. B* **126** (1977) 298. doi:10.1016/0550-3213(77)90384-4.

[19] J. Pumplin, D. Stump, J. Huston et al., "New Generation of Parton Distributions with Uncertainties from Global QCD Analysis.", *JHEP* **0207** (2002) 012. doi:10.1088/1126-6708/2002/07/012.

[20] A. Martin, W. Stirling, R. Thorne et al., "Parton distributions for the LHC", *Eur. Phys. J. C* **63** (2009) 189–285. 10.1140/epjc/s10052-009-1072-5.

[21] CMS Collaboration, "CMS Physics Technical Design Report Volume II: Physics Performance", *J. Phys. G* **34** (2006), no. CERN-LHCC-2006-021. CMS-TDR-008-2, 995–1579. doi:10.1088/0954-3899/34/6/S01.

[22] C. Anastasiou, L. Dixon, K. Melnikov et al., "High-precision QCD at hadron colliders: Electroweak gauge boson rapidity distributions at next-to-next-to leading order", *Phys. Rev. D* **69** (May, 2004) 094008. doi:10.1103/PhysRevD.69.094008.

[23] P. J. Rijken and W. L. van Neerven, "Order α_s^2 contributions to the Drell-Yan cross section at fixed target energies", *Phys. Rev. D* **51** (Jan, 1995) 44–63. doi:10.1103/PhysRevD.51.44.

[24] R. Hamberg, W. van Neerven, and T. Matsuura, "A complete calculation of the order α_s^2 correction to the Drell-Yan K-factor", *Nuclear Physics B* **359** (1991), no. 23, 343 – 405. doi:10.1016/0550-3213(91)90064-5.

[25] W. van Neerven and E. Zijlstra, "The O(α_s^2) corrected Drell-Yan K-factor in the DIS and MS schemes", *Nuclear Physics B* **382** (1992), no. 1, 11 – 62. doi:10.1016/0550-3213(92)90078-P.

[26] T. Sjostrand, S. Mrenna, and P. Z. Skands, "PYTHIA 6.4 Physics and Manual", *JHEP* **05** (2006) 026, arXiv:hep-ph/0603175. doi:10.1088/1126-6708/2006/05/026.

[27] G. Miu and T. Sjöstrand *Phys. Lett.* **B449** (1999) 313.

[28] M. Bengtsson and T. Sjöstrand *Phys. Lett.* **B185** (1987) 435.

[29] T. Sjöstrand and P. Skands *Eur. Phys. J. C* **39** (2005) 129.

[30] B. Andersson, G. Gustafson, G. Ingelman et al. *Phys. Rep.* **97** (1983) 31.

[31] B. Andersson, "The Lund Model". Cambridge University Press, 1998.

[32] B. Webber *Nucl. Phys.* **B238** (1984) 492.

[33] TOTEM Collaboration, "The TOTEM experiment at the CERN LHC", *JINST* **0803** (2008) S08007. doi:10.1088/1748-0221/3/08/S08007.

[34] TOTEM Collaboration, "First measurement of the total proton-proton cross section at the LHC energy of $\sqrt{s} = 7$ TeV", *EPL* **96** (2011) 21002. doi:10.1209/0295-5075/96/21002.

[35] CMS Collaboration, "Measurement of the inelastic pp cross section at $\sqrt{s} = 7$ TeV with the CMS detector.", *CMS-PAS-FWD-11-001* (2011).

[36] The CMS and TOTEM diffractive and forward physics working group, "Prospects for Diffractive and Forward Physics at the LHC", *CERN-LHCC-2006-039-G-124* (2006).

[37] P. Collins, "An Introduction to Regge Theory and High-Energy Physics". Cambridge University Press, 1977.

[38] UA8 Collaboration, "Measurements of single diffraction at $\sqrt{s} = 630$ GeV: Evidence for a non-linear $\alpha(t)$ of the pomeron", *Nucl. Phys.* **B514** (1998) 3. doi:10.1016/S0550-3213(97)00813-4.

[39] ZEUS Collaboration, "A QCD analysis of ZEUS diffractive data", *Nucl. Phys. B* **831** (2010) 1, arXiv:0911.4119. doi:10.1016/j.nuclphysb.2010.01.014.

[40] H1 Collaboration, "Dijet Cross Sections and Parton Densities in Diffractive DIS at HERA", *JHEP* **0710** (2007) 042, arXiv:0708.3217. doi:10.1088/1126-6708/2007/10/042.

[41] H1 Collaboration, "Measurement and QCD analysis of the diffractive deep-inelastic scattering cross-section at HERA", *Eur. Phys. J. C* **48** (2006) 715, arXiv:hep-ex/0606004. doi:10.1140/epjc/s10052-006-0035-3.

[42] CDF Collaboration, "Observation of diffractive W boson production at the Tevatron", *Phys. Rev. Lett.* **78** (1997) 2698, arXiv:hep-ex/9703010. doi:10.1103/PhysRevLett.78.2698.

[43] D0 Collaboration, "Observation of diffractively produced W and Z bosons in $\bar{p}p$ collisions at $\sqrt{s} = 1800$ GeV", *Phys. Lett. B* **574** (2003) 169, arXiv:hep-ex/0308032. doi:10.1016/j.physletb.2003.09.001.

[44] J. C. Collins, "Proof of factorization for diffractive hard scattering", *Phys. Rev. D* **57** (1998) 3051, arXiv:hep-ph/9709499. doi:10.1103/PhysRevD.57.3051.

[45] J. D. Bjorken, "Rapidity gaps and jets as a new physics signature in very high-energy hadron hadron collisions", *Phys. Rev. D* **47** (1993) 101. doi:10.1103/PhysRevD.47.101.

[46] A. B. Kaidalov, V. A. Khoze, A. D. Martin et al., "Probabilities of rapidity gaps in high-energy interactions", *Eur. Phys. J. D* **21** (2001) 521, arXiv:hep-ph/0105145. doi:10.1007/s100520100751.

[47] CDF Collaboration, "Diffractive W and Z Production at the Fermilab Tevatron", *Phys. Rev. D* **82** (2010) 112004, arXiv:1007.5048. doi:10.1103/PhysRevD.82.112004.

[48] GEANT4 Collaboration, "GEANT4: A simulation toolkit", *Nucl. Instrum. Meth.* **A506** (2003) 250.

[49] P. M. Nadolsky et al., "Implications of CTEQ global analysis for collider observables", *Phys. Rev. D* **78** (2008) 013004, arXiv:0802.0007. doi:10.1103/PhysRevD.78.013004.

[50] T. Sjostrand, S. Mrenna, and P. Skands, "A Brief Introduction to PYTHIA 8.1", *Comp. Phys. Com.* **178** (2008) arXiv:0710.3820. doi:10.1016/j.cpc.2008.01.036.

[51] P. Bartalini and L. Fanó, eds., "Multiple Parton Interactions at the LHC. Proceedings, 1st Workshop, Perugia, Italy, October 27-31, 2008". (2009). DESY-PROC-2009-06.

[52] R. Field, "Studying the underlying event at CDF and the LHC", in *Proceedings of the First International Workshop on Multiple Partonic Interactions at the LHC MPI'08, October 27-31, 2008*, P. Bartalini and L. Fanó, eds., pp. 12–31. Perugia, Italy, October, 2009. arXiv:1003.4220.

[53] A. Buckley et al., "Systematic event generator tuning for the LHC", *Eur. Phys. J. C* **65** (2010) 331, arXiv:0907.2973. doi:10.1140/epjc/s10052-009-1196-7.

[54] P. Z. Skands, "Tuning Monte Carlo Generators: The Perugia Tunes", *Phys. Rev. D* **82** (2010) 074018, arXiv:1005.3457. doi:10.1103/PhysRevD.82.074018.

[55] R. Field, "Early LHC Underlying Event Data - Findings and Surprises", (2010). arXiv:1010.3558.

[56] R. Corke and T. Sjöstrand, "Interleaved Parton Showers and Tuning Prospects", *JHEP* **03** (2011) 032, arXiv:1011.1759. doi:10.1007/JHEP03(2011)032.

[57] CMS Collaboation, "Measurement of the underlying event activity at the LHC with $\sqrt{s} = 7\,\text{TeV}$ and comparison with $\sqrt{s} = 0.9\,\text{TeV}$.", *Journal of High Energy Physics* **2011** (2011) 1–31. doi:10.1007/JHEP09(2011)109.

[58] P. Bruni, A. Edin, and G. Ingelman. http://www3.tsl.uu.se/thep/MC/pompyt/.

[59] P. Bruni and G. Ingelman, "Diffractive W and Z production at $p\bar{p}$ colliders and the pomeron parton content", *Phys. Lett. B* **311** (1993) 317. doi:10.1016/0370-2693(93)90576-4.

[60] P. Bruni and G. Ingelman, "Diffractive hard scattering at e p and $p\bar{p}$ colliders", in *Proceedings of the International Europhysics Conference on High Energy Physics*, p. 595. 1994. Marseille, France, 22-28 Jul 1993.

[61] H1 Collaboration, "Dijet Cross Sections and Parton Densities in Diffractive DIS at HERA", *JHEP* **10** (2007) 042, arXiv:0708.3217. doi:10.1088/1126-6708/2007/10/042.

[62] T. Stelzer and W. Long, "Automatic generation of tree level helicity amplitudes.", *Computer Physics Communications* **81** (1994) 357.

[63] F. Maltoni and T. Stelzer, "MadEvent: automatic event generation with MadGraph", *J. High Energy Phys.* **02** (2003) 027.

[64] MadGraph Team. https://server06.fynu.ucl.ac.be/projects/madgraph/wiki/MadGraphSamples.

[65] O. Bruening et al., "LHC Design Report Vol. I: The LHC Main Ring", *CERN-2004-003-V-1* (2004).

[66] W. J. Stirling. http://mstwpdf.hepforge.org/.

[67] CERN. http://cdsweb.cern.ch/record/1260465.

[68] CMS Collaboration, "The CMS experiment at the CERN LHC", *JINST* **0803** (2008) S08004. `doi:10.1088/1748-0221/3/08/S08004`.

[69] ATLAS Collaboration, "Detector and Physics Performance Technical Design Report. Volume 1", *CERN-LHCC-99-14* (1999).

[70] LHCb Collaboration, "LHCb technical proposal", *CERN-LHCC-98-04* (1998).

[71] ALICE Collaboration, "Technical Proposal for a Large Ion Collider Experiment at the CERN LHC", *CERN-LHCC-95-71* (1995).

[72] CMS. http://cdsweb.cern.ch/record/1433717.

[73] CMS. http://cms.web.cern.ch/news/detector-overview.

[74] CMS Collaboration, "CMS Physics Technical Design Report Volume I: Detector Performance and Software", *CERN-LHCC-2006-001* (2006).

[75] CMS Collaboration, "The CMS High Level Trigger", *Eur. Phys. J. C* **46** (2005) 605–667. `doi:10.1140/epjc/s2006-02495-8`.

[76] CMS. https://twiki.cern.ch/twiki/bin/view/CMSPublic/PhysicsResults.

[77] CMS Collaboration, "Measurement of CMS Luminosity", *CMS-PAS-EWK-10-004* (2010).

[78] CMS. https://twiki.cern.ch/twiki/bin/view/CMSPublic/LumiPublicResults2010.

[79] CMS. https://twiki.cern.ch/twiki/bin/view/CMSPublic/LumiPublicResults2011.

[80] "LHC Performance Workshop 2011", *CERN-ATS-2011-005* (2011).

[81] "LHC Performance Workshop 2012".
https://indico.cern.ch/conferenceDisplay.py?confId=170230, 2012.

[82] M. Dittmar, F. Pauss, and D. Zürcher, "Towards a precise parton luminosity determination at the CERN LHC", *Phys. Rev. D* **56** (Dec, 1997) 7284–7290. doi:10.1103/PhysRevD.56.7284.

[83] CMS Collaboration, "Measurement of inclusive W and Z cross sections in pp collisions at $\sqrt{s} = 7\,\text{TeV}$.", *J. High Energy Phys.* **1** (2011) 080. doi:10.1007/JHEP01(2011)080.

[84] CMS Collaboration, "Measurement of the inclusive W and Z production cross sections in pp collisions at $\sqrt{s} = 7\,\text{TeV}$ with the CMS experiment.", *J. High Energy Phys.* **10** (2011) 132. doi:10.1007/JHEP10(2011)132.

[85] ALEPH, DELPHI, L3, OPAL, SLD, LEP Electroweak Working Group, SLD Electroweak Group and SLD Heavy Flavour Group collaborations, "Precision electroweak measurements on the Z resonance", *Phys. Rept.* **427** (2006) 257. doi:10.1016/j.physrep.2005.12.006.

[86] ATLAS Collaboration, "Measurement of the $W \to l\nu$ and $Z/\gamma* \to ll$ production cross sections in proton-proton collisions at $\sqrt{s} = 7\,\text{TeV}$ with the ATLAS detector.", *J. High Energy Phys.* **12** (2010) 060. doi:10.1007/JHEP12(2010)060.

[87] S. D. Drell and T.-M. Yan, "Massive Lepton-Pair Production in Hadron-Hadron Collisions at High Energies", *Phys. Rev. Lett.* **25** (Aug, 1970) 316–320. doi:10.1103/PhysRevLett.25.316.

[88] CMS Collaboration, "Measurement of the Drell-Yan cross section in pp collisions at $\sqrt{s} = 7\,\text{TeV}$.", *J. High Energy Phys.* **10** (2011) 007. doi:10.1007/JHEP10(2011)007.

[89] CMS Collaboration, "Measurement of the rapidity and transverse momentum distributions of Z bosons in pp collisions at $\sqrt{(s)}=7$ TeV", *Phys. Rev. D* **85** (Feb, 2012) 032002. doi:10.1103/PhysRevD.85.032002.

[90] CMS Collaboration, "Measurement of the lepton charge asymmetry in inclusive W production in pp collisions at $\sqrt{s} = 7\,\text{TeV}$.", *J. High Energy Phys.* **04** (2011) 050. doi:10.1007/JHEP04(2011)050.

[91] CMS Collaboration, "Measurement of the muon charge asymmetry in inclusive W production in pp collisions at $\sqrt{s} = 7\,\text{TeV}$.", *CMS-PAS-EWK-11-005*. (2011).

[92] CMS Collaboration, "Measurement of the electron charge asymmetry in inclusive W production in pp collisions at $\sqrt{s} = 7\,\text{TeV}$.", *CMS-PAS-SMP-12-001*. (2012).

[93] C. F. Berger, Z. Bern, L. J. Dixon et al., "Next-to-leading order QCD predictions for W + 3-jet distributions at hadron colliders", *Phys. Rev. D* **80** (Oct, 2009) 074036. doi:10.1103/PhysRevD.80.074036.

[94] Z. Bern, G. Diana, L. J. Dixon et al., "Left-handed W bosons at the LHC", *Phys. Rev. D* **84** (Aug, 2011) 034008. doi:10.1103/PhysRevD.84.034008.

[95] CMS Collaboration, "Measurement of the Polarization of W Bosons with Large Transverse Momenta in W + jets Events at the LHC", *Phys. Rev. Lett.* **107** (Jul, 2011) 021802. doi:10.1103/PhysRevLett.107.021802.

[96] CMS Collaboation, "First measurement of the underlying event activity at the LHC with $\sqrt{s} = 0.9\,\text{TeV}$.", *Eur. Phys. J. C* **70** (2010) 555. doi:10.1140/epjc/s10052-010-1453-9.

[97] CMS Collaboration, "Study of the Underlying Event at Forward Rapidity in Proton-Proton Collisions at the LHC", *CMS-PAS-FWD-11-003* (2012).

[98] CMS Collaboration, "Charged particle multiplicitites in pp interactions at $\sqrt{s} = 0.9, 2.36,$ and $7\,\text{TeV}$", *J. High Energy Phys.* **01** (2010) 079. doi:10.1007/JHEP01(2011)079.

[99] CMS Collaboration, "Transverse-Momentum and Pseudorapidity Distributions of Charged Hadrons in pp Collisions at $\sqrt{s} = 0.9 \text{and} 2.36 \text{TeV}$", *J. High Energy Phys.* **02** (2010) 041. doi:10.1007/JHEP02(2010)041.

[100] CMS Collaboration, "Transverse-Momentum and Pseudorapidity Distributions of Charged Hadrons in pp Collisions at $\sqrt{s} = 7\,\text{TeV}$", *Phys. Rev. Lett.* **105** (Jul, 2010) 022002. doi:10.1103/PhysRevLett.105.022002.

[101] CMS Collaboration, "Charged particle transverse momentum spectra in pp collisions at $\sqrt{s} = 0.9$ and $7\,\text{TeV}$", *J. High Energy Phys.* **08** (2011) 086. doi:10.1007/JHEP08(2011)086.

[102] CMS Collaboration, "Pseudorapidity distributions of charged particles in pp collisions at $\sqrt{s} = 7\,\text{TeV}$ with at least one central charged particles", *CMS-PAS-QCD-10-024* (2011).

[103] CMS. https://twiki.cern.ch/twiki/bin/view/CMSPublic/PhysicsResultsFSQ.

[104] T. Sjöstrand and M. Van Zijl, "Multiple parton-parton interactions in an impact parameter picture", *Physics Letters B* **188** (1987), no. 1, 149 – 154. `doi:10.1016/0370-2693(87)90722-2`.

[105] L. Frankfurt, M. Strikman, and C. Weiss, "Transverse nucleon structure and diagnostics of hard parton-parton processes at LHC", *Phys. Rev. D* **83** (Mar, 2011) 054012. `doi:10.1103/PhysRevD.83.054012`.

[106] CMS Collaboration, "Measurement of the underlying event in the Drell-Yan process in proton-proton collisions at $\sqrt{s}$ =7 TeV", (2012). `arXiv:1204.1411`. Submitted to EJPC.

[107] R. Gluckstern, "Uncertainties in track momentum and direction, due to multiple scattering and measurement errors", *Nuclear Instruments and Methods* **24** (1963), no. 0, 381 – 389. `doi:10.1016/0029-554X(63)90347-1`.

[108] CMS Collaboration, "Measurement of Momentum Scale and Resolution using Low-mass Resonances and Cosmic Ray Muons", *CMS-PAS-TRK-10-004* (2010).

[109] CMS Collaboration, "Missing E_T Performance in CMS.", *CMS-PAS-JME-07-001* (2007).

[110] CMS Collaboration, "Performance of Track-Corrected Missing E_T in CMS.", *CMS-PAS-JME-09-010* (2009).

[111] CMS Collaboration, "Particle-Flow Event Reconstruction in CMS.", *CMS-PAS-PFT-09-001* (2009).

[112] CMS Collaboration, "Missing Transverse Energy Performance in Minimum-Bias and Jet Events from Proton-Proton Collisions at $\sqrt{s} = 7$ TeV.", *CMS-PAS-JME-10-004* (2010).

[113] CMS Collaboration, "CMS MET Performance in Events Containing Electroweak Bosons from pp Collisions at $\sqrt{s} = 7$ TeV.", *CMS-PAS-JME-10-005* (2010).

[114] M. Cacciari, G. Salam, and G. Soyez, "The anti-k_t jet clustering algorithm", *JHEP* **04** (2008) 063. `doi:10.1088/1126-6708/2008/04/063`.

[115] CMS Collaboration, "Jet Performance in pp Collisions at $\sqrt{s} = 7$ TeV", *CMS-PAS-JME-10-003* (2010).

[116] CMS Collaboration, "Jet Energy Resolution in CMS at $\sqrt{s} = 7$ TeV", *CMS-PAS-JME-10-014* (2011).

[117] CMS Collaboration, "Determination of the Jet Energy Scale in CMS with pp Collisions at $\sqrt{s} = 7$ TeV", *CMS-PAS-JME-10-010* (2010).

[118] CMS Collaboration, "CMS Tracking Performance Results from early LHC Operation", *Eur. Phys. J. C* **70** (2010) 1165, arXiv:1007.1988. doi:10.1140/epjc/s10052-010-1491-3.

[119] CMS Collaboration, "Tracking and Vertexing Results from First Collisions", *CMS-PAS-TRK-10-001* (2010).

[120] ATLAS Collaboration, "Measurement of the Inelastic Proton-Proton Cross-Section at $\sqrt{s} = 7$ TeV with the ATLAS Detector", (2011). arXiv:1104.0326.

[121] TOTEM Collaboration, "Proton-proton elastic scattering at the LHC energy of $\sqrt{s} = 7$ TeV", *Europhys. Lett.* **95** (2011) 41001. doi:10.1209/0295-5075/95/41001.

[122] CMS Collaboration, "Commissioning of the Particle-Flow Reconstruction in Minimum-Bias and Jet Events from pp Collisions at 7 TeV", CMS Physics Analysis Summary CMS-PAS-PFT-10-002, (2010).

[123] H. Jung. Thanks to Hannes for the figure.

[124] C. Anastasiou, L. Dixon, K. Melnikov et al., "High-precision QCD at hadron colliders: Electroweak gauge boson rapidity distributions at next-to-next-to leading order", *Phys. Rev. D* **69** (May, 2004) 094008. doi:10.1103/PhysRevD.69.094008.

[125] K. Melnikov and F. Petriello, "Electroweak gauge boson production at hadron colliders through $O(\alpha_s^2)$", *Phys. Rev. D* **69** (Dec, 2006) 114017. doi:10.1103/PhysRevD.74.114017.

[126] P. M. Nadolsky, H.-L. Lai, Q.-H. Cao et al., "Implications of CTEQ global analysis for collider observables", *Phys.Rev.* **D78** (2008) 013004, arXiv:0802.0007. doi:10.1103/PhysRevD.78.013004.

Acknowledgments

For this experience of working in such a large and bustling collaboration to turn out so fruitful, inspiring, and pleasant I am deeply indebted and thankful to so many.

First, I want to thank CERN and the CMS Publication Committee Chair for permission to include material with CERN copyright.

I would like to express my gratitude to Professor Günther Dissertori for giving me the opportunity to be part of it all. For all his support in these years and for being a role model performing always to perfection, encouraging to push my own limits forward again and again.

From Michael Dittmar I learned not only the necessary skills to carry out this thesis but, more importantly, what it means to question, to do real science. And to look further than work. It has been wonderfully inspiring to see his involvement for a better world with such sincere commitment over all these years. Thank you, Michael, for these valuable lessons.

I would like to thank Carmelo Marchica for introducing me with heroic patience to the complex world of CMS and for the many hours we spent over extensive physics discussions. And for friendship. A special thanks to Jürg Eugster for being so easy to work with, never trying to show-off but instead getting the work done. Not only the work has been pleasant, the breaks in between, to have somebody to share my dreams and passion for the mountains, were rewarding.

I am indebted to Hannes Jung who was ceaselessly standing by our side, supporting us with his immense knowledge and good advises. I also thank him and Rainer Wallny for reading this thesis and accepting on being co-examiners.

The reliable support in so manifold ways of Gabrielle Kogler, Peter Meierhofer, Johanna Amberg and Nadja Sigrist are warmly acknowledged. I would particularly like to thank Gabrielle for caring for us and for being so kind.

I had a wonderful time with my colleagues from the IPP, those in Geneva and those in Zurich. I want to thank them all for the good time, for always being friendly and supportive. Thanks to Pascal, Pedja, Christina, and Jürg for the great laughs, discussions and fun we had, and especially for friendship.

Above all my profoundest thanks to my incomparable family. To my parents, who knew how to raise me to pursue my own goals whatever the circumstances. To Olinda for being there as best friend, with such an astonishing talent at listening and advising. And to Tommy, for making every single day of these past years a fulfilled and happy one. This thesis is dedicated to them.

I want morebooks!

Buy your books fast and straightforward online - at one of the world's fastest growing online book stores! Environmentally sound due to Print-on-Demand technologies.

Buy your books online at
www.get-morebooks.com

Kaufen Sie Ihre Bücher schnell und unkompliziert online – auf einer der am schnellsten wachsenden Buchhandelsplattformen weltweit!
Dank Print-On-Demand umwelt- und ressourcenschonend produziert.

Bücher schneller online kaufen
www.morebooks.de

OmniScriptum Marketing DEU GmbH
Heinrich-Böcking-Str. 6-8
D - 66121 Saarbrücken
Telefax: +49 681 93 81 567-9

info@omniscriptum.com
www.omniscriptum.com

Printed by Books on Demand GmbH, Norderstedt / Germany